The Energy Balance
in Northeast Asia

JOSEPH A. YAGER
with the assistance of Shelley M. Matsuba

The Energy Balance
in Northeast Asia

THE BROOKINGS INSTITUTION
Washington, D.C.

Copyright © 1984 by
THE BROOKINGS INSTITUTION
1775 Massachusetts Avenue, N.W., Washington, D.C. 20036

Library of Congress Cataloging in Publication data:

Yager, Joseph A., 1916–
   The energy balance in Northeast Asia.

   Includes index.
   1. Power resources—Economic aspects—East Asia.
2. Power resources—Political aspects—East Asia.
3. Energy policy—East Asia. I. Matsuba, Shelley M.,
1953–   . II. Brookings Institution. III. Title.
HD9502.A782Y33 1984   333.79′095   84-45276
ISBN 0-8157-9672-2
ISBN 0-8157-9671-4 (pbk.)

1 2 3 4 5 6 7 8 9

THE BROOKINGS INSTITUTION is an independent organization devoted to nonpartisan research, education, and publication in economics, government, foreign policy, and the social sciences generally. Its principal purposes are to aid in the development of sound public policies and to promote public understanding of issues of national importance.

The Institution was founded on December 8, 1927, to merge the activities of the Institute for Government Research, founded in 1916, the Institute of Economics, founded in 1922, and the Robert Brookings Graduate School of Economics and Government, founded in 1924.

The Board of Trustees is responsible for the general administration of the Institution, while the immediate direction of the policies, program, and staff is vested in the President, assisted by an advisory committee of the officers and staff. The by-laws of the Institution state: "It is the function of the Trustees to make possible the conduct of scientific research, and publication, under the most favorable conditions, and to safeguard the independence of the research staff in the pursuit of their studies and in the publication of the results of such studies. It is not a part of their function to determine, control, or influence the conduct of particular investigations or the conclusions reached."

The President bears final responsibility for the decision to publish a manuscript as a Brookings book. In reaching his judgment on the competence, accuracy, and objectivity of each study, the President is advised by the director of the appropriate research program and weighs the views of a panel of expert outside readers who report to him in confidence on the quality of the work. Publication of a work signifies that it is deemed a competent treatment worthy of public consideration but does not imply endorsement of conclusions or recommendations.

The Institution maintains its position of neutrality on issues of public policy in order to safeguard the intellectual freedom of the staff. Hence interpretations or conclusions in Brookings publications should be understood to be solely those of the authors and should not be attributed to the Institution, to its trustees, officers, or other staff members, or to the organizations that support its research.

Foreword

THE economies of non-Communist Northeast Asia—Japan, South Korea, and Taiwan—have displayed unusual dynamism. Their growth rates have been among the highest in the world. These economies are, however, poor in energy resources. Their expansion has been possible only through increased reliance on imported fuels. Japan now imports more energy than any other country; South Korea and Taiwan also are major energy importers. Not only could knowledge of past energy developments in Northeast Asia be useful in analyzing the problems of other energy-importing areas, but future energy developments in that region could also have important effects on international energy markets and on international relations in general.

In this book Joseph A. Yager and Shelley M. Matsuba explain the changes in total energy consumption and patterns of energy supply and use that occurred in Northeast Asia during the 1960s and 1970s. They compare the energy experiences of South Korea and Taiwan—two of the world's most advanced developing countries—and the energy experience of Japan with those of three other industrialized countries. Special attention is paid to the adjustment to oil crises and to the relation between energy security and foreign policy.

The authors also project possible future energy requirements in Northeast Asia under different assumptions. They estimate the quantities of fuels—oil, coal, natural gas, and uranium—that would be needed to meet those requirements and suggest where those fuels would be obtained. These estimates are necessarily only illustrative, but the process by which they are derived brings out the complex interactions that will determine future energy developments in the region. The appendix defines the technical terms used in the study and explains the authors' methods of analysis.

This book was written as part of the Brookings Foreign Policy Studies program directed by John D. Steinbruner. Joseph A. Yager was a Brookings senior fellow when the book was written and is now a guest scholar at Brookings. Shelley M. Matsuba was on the research staff of Resources for the Future when she worked on the book and has since joined Inslaw, Inc.

The authors express their appreciation to Edward Allen, Joy Dunkerley, John E. Gray, and Lee Schipper for their comments on the manuscript. The authors also benefited from the generous assistance and advice of a number of people in Northeast Asia. They are particularly indebted to L. K. Chen, T. T. Feng, Ho Tak Kim, Hoesung Lee, Kenichi Matsui, Chen Sun, Sung Nack Chung, and Takao Tomitate. They also thank Ann Ziegler, who served as project secretary; James McEuen, who edited the manuscript; Alan G. Hoden, who checked it for accuracy; and Ward & Silvan, who prepared the index. Brookings is grateful to the U.S. Central Intelligence Agency for financial support of this project.

The views in this book are those of the authors and should not be ascribed to the persons or organizations whose assistance is acknowledged above, or to the trustees, officers, or staff members of the Brookings Institution.

BRUCE K. MACLAURY
President

July 1984
Washington, D.C.

Contents

ix

Tables

CHAPTER ONE

Overview

BEFORE the oil crisis of 1973–74, the economies of the non-Communist countries of Northeast Asia—Japan, the Republic of Korea (hereafter South Korea or Korea), and the Republic of China (hereafter Taiwan)—were growing rapidly.[1] To meet their rising energy requirements, they were forced to import increasing amounts of energy, largely in the form of convenient, readily available, and relatively cheap crude oil. In 1973, on the eve of the first oil crisis, imported oil provided 78 percent of Japan's energy, 53 percent of South Korea's, and 67 percent of Taiwan's.

In the years before the 1973–74 oil crisis, the overall energy intensity of the Japanese economy was increasing. (That is, the amount of energy needed to produce a unit of gross domestic product, GDP, was rising.) During the same period the energy intensity of the Korean economy decreased slightly. The energy intensity of Taiwan's economy fluctuated but was almost the same in 1973 as it had been in 1960.

The rise in the energy intensity of the Japanese economy in the 1960s and early 1970s can be explained largely by falling energy prices (in real terms) and by the increased importance of manufacturing in the economy. The decline in the energy intensity of the Korean economy can be attributed to a marked increase in the thermal efficiency of energy use made possible principally by a shift to oil from less efficient coal, wood, charcoal, and crop residues. The improvement in thermal efficiency overcame the effects of changes, such as the growing importance of manufacturing, that tended to push up energy intensity. In Taiwan, an increase in thermal efficiency was almost exactly balanced by opposite

1. South Korea and Taiwan are generally regarded as parts of Korea and China, respectively. They are referred to here as "countries" only in a geographic sense.

1

factors, including the increased importance of manufacturing and larger energy losses in generating electricity.

The 1973–74 oil crisis, with its fourfold rise in international oil prices, was an economic and energy turning point in Japan but not in Korea. The effect of the crisis in Taiwan was sharp but brief. In each of the three countries, the adjustment to the huge increase in oil prices was strongly influenced by the state of the economy when the crisis struck and by the choice that the government made between the partly conflicting goals of price stability and economic growth.

In the years immediately preceding the 1973–74 crisis, Japan's GDP had been growing at an average annual rate of 9 percent, and its energy consumption had been increasing at a rate of over 10 percent a year. The jump in oil prices and the measures taken by the government to check inflation brought economic growth to a virtual standstill in 1974 and 1975. Energy consumption declined in both of these years. The Japanese economy recovered in 1976 and subsequent years, but—in contrast with the experience of the United States, France, and the Federal Republic of Germany (hereafter, West Germany or Germany)—did not even come close to regaining its rate of growth before the crisis. Energy consumption after the crisis, however, grew only about half as fast as did GDP, so that the overall energy intensity of the economy declined. The principal cause of this decline was a fall in the intensity of energy use in manufacturing.

In Korea, the government appears deliberately to have accepted an increase in the rate of inflation in order not to interfere with continued rapid economic growth. GDP actually grew somewhat more rapidly after the oil crisis than before. The gentle decline in energy intensity continued, propelled by higher real energy prices, continued increases in thermal efficiency, and larger net imports of embodied energy in the country's nonfuel international trade.

The government of the Republic of China on Taiwan, like the government of Japan, took strong measures to counter the inflationary impact of the rise in international oil prices. The immediate effects of these measures on the growth of GDP and energy consumption were similar to those produced in Japan, although somewhat less severe. In Taiwan, however, both GDP and energy consumption rebounded sharply. GDP did not quite regain the growth rate of the years before the oil crisis, but energy consumption grew somewhat more rapidly. The resulting moderate increase in energy intensity can be attributed to increased energy losses in generating electricity, larger net exports of embodied energy, and declining real energy prices after 1974.

After the 1973–74 crisis, all three Northeast Asian countries had strong incentives to reduce their dependence on Middle Eastern oil. But finding oil supplies outside the Middle East proved to be difficult; substituting other sources of energy for oil was more feasible. Japan took the lead in the substitution strategy. Primarily by increasing its reliance on nuclear energy and liquefied natural gas, Japan reduced the share of oil in its total energy supply from 76.1 percent in 1973 to 71.5 percent in 1979.

In contrast, the share of oil in the total energy supplies of Korea and Taiwan increased from 1973 to 1979 (from 53.9 percent to 61.0 percent in Korea, and from 68.9 percent to 72.3 percent in Taiwan). The dependence of these countries on oil would have been even greater if they had not increased their coal imports and begun to produce some electricity in nuclear power plants.

The second oil crisis, in 1979–80, caused international oil prices to double and gave oil-importing countries additional incentives to conserve energy and to substitute other fuels for oil. The Japanese economy was especially well prepared to absorb the shock of the second oil crisis. The first crisis had induced a decrease in the energy intensity of the manufacturing sector. Prices were relatively stable, labor productivity was rising faster than wages, and excess productive capacity existed. The Japanese government was able to bring about a smooth adjustment to the increase in oil prices simply by reinforcing natural economic forces with mildly deflationary fiscal and monetary policies. The governments of Korea and Taiwan also achieved fairly successful adjustments to the second oil crisis through policies that struck a balance between the goals of price stability and economic growth.

The adjustments of the three Northeast Asian economies to the second oil crisis had not fully run their course when a worldwide recession caused a weakening in the demand for oil and a decline in international oil prices. Recent years, therefore, cannot be regarded as having established any new trends either in energy markets or in the energy economies of individual countries. The illustrative projections of possible future developments in Northeast Asia presented in this study do not draw much on events since 1980. Rather, they are based principally on trends, relationships, and—to some extent—plans that emerged in the years between the two oil crises.

The central case of this study, which assumes stable energy prices and specified economic growth rates, projects a 50 percent increase in the primary energy requirements of the Northeast Asian countries from 1980 to 1990.[2] The energy requirements of individual Northeast Asian countries are projected to expand more rapidly than the requirements of comparable countries in other parts of the world. Thus the share of Northeast Asia in international energy markets is also likely to expand.

The structure of energy supply in Northeast Asia will continue to change during the 1980s. The share of oil is projected to decline from 65 percent in 1980 to 57 percent in 1990. The share of nuclear energy is projected to double, going from 5 percent to 10 percent of total energy supply. Small increases are also projected for the shares of natural gas and coal. Absolute import requirements for all fuels are estimated to be higher in 1990 than in 1980. Despite efforts to reduce dependence on imported oil, regional import requirements for oil are projected to rise about 30 percent.

Energy requirements of the three Northeast Asian countries after 1990 will depend in general on the rates of economic growth and the energy intensities of these economies. The shift to technology-intensive activities that is already under way could continue and could help to sustain both respectable rates of growth and a gradual decline in energy intensities. It is unlikely, however, that the rates of decline in energy intensity will be so great as to cancel the positive effect of economic growth on energy requirements. The move away from heavy reliance on oil may slow in the 1990s as economically attractive opportunities to substitute other fuels for oil are gradually exhausted.

Two conclusions of wider significance can be drawn from the analysis of past energy developments in Northeast Asia. First, in adjusting to large increases in energy prices, governments must choose between the partly competing goals of economic growth and price stability. Second, governments cannot control all of the factors that determine the overall energy intensity of an economy. When governments do influence energy intensity, they often do so inadvertently or by a conscious decision to give priority to goals other than energy conservation.

Possible energy developments in Northeast Asia have several impor-

2. The use of 1990 in this projection should not be taken too literally. Recent developments in Japan—slower economic growth and reduced energy intensity—suggest that the projected level of primary energy requirements may not be reached until sometime after 1990.

tant implications. The growing importance of the region in international energy markets will cause Northeast Asian firms and government agencies to become even more involved than they are now in energy exploration and development projects in other parts of the world. At the same time, producers of energy-intensive commodities and energy-related equipment and services will probably find expanding markets in Northeast Asia.

To pay for rising energy imports, the Northeast Asian countries must either borrow abroad or generate increasing export surpluses in their nonfuel trade. Only Korea has relied on borrowing in the past, and in the future it will probably join Japan and Taiwan in paying for energy imports from current foreign exchange earnings. In running growing export surpluses in their nonfuel trade, the Northeast Asian countries will inevitably compete with industrialized countries that export similar commodities.

In shifting from primary dependence on Middle Eastern oil to reliance on a variety of energy sources from Pacific countries—coal from Australia, Canada, and the United States; liquefied natural gas from Southeast Asia and Alaska; oil from Southeast Asia; uranium from Canada and Australia; and uranium enrichment services from the United States—the Northeast Asian countries are adding substance to the concept of a Pacific Basin community.

The expansion of civil nuclear energy facilities in Northeast Asia will create problems, as well as opportunities, for regional cooperation. In Northeast Asia—in contrast with recent U.S. experience—nuclear electricity is cheap electricity. As more nuclear electricity becomes available, the competitiveness of some Northeast Asian exports will improve. Perhaps more important, the expanding Northeast Asian nuclear energy programs could increase the risk of further proliferation of nuclear weapons. Spent nuclear fuel from power reactors contains plutonium that can either be recycled as fuel or used to make nuclear explosives. Some spent fuel from Japanese power reactors has been reprocessed in Japan and in Western Europe to extract plutonium and residual uranium for use as fuel. Korea and Taiwan will also want to reprocess spent fuel or to have it reprocessed elsewhere. Devising better controls for separated plutonium will become increasingly important. Additional storage space will also be needed for spent fuel before it is reprocessed or disposed of permanently. Regional approaches to these problems may prove to be more desirable than national ones.

CHAPTER TWO

Energy Developments in Japan in the 1960s and 1970s

THE 1960s and 1970s were years of rapid economic change in Japan. The gross domestic product measured in constant yen more than quadrupled. Population grew at an average annual rate of only 1.1 percent, and GDP per capita in 1979 was over three and a half times what it had been in 1960 (table 2-1). In contrast with Japan's remarkable economic achievement, the GDPs of the United States and of the European Community roughly doubled in the same period.

The structure of Japan's economy also changed greatly, as a comparison of the percentage contributions to GDP of principal economic sectors in 1960 and 1979 shows:[1]

	1960	1979
Agriculture, forestry, and fishing	14.3	4.4
Manufacturing	26.8	34.7
Commerce	36.1	39.1
Transportation, storage, and communications	6.1	5.7
Government services	10.8	8.3
Other	5.9	7.8

The greatest structural changes were the marked increase in the share of manufacturing and the sharp decline in the importance of agriculture, forestry, and fishing. In absolute terms, value added in manufacturing

1. Organization for Economic Cooperation and Development, personal communication to author for 1960 data; and *National Accounts of OECD Countries, 1962–79*, vol. 2 (Paris: OECD, 1981), p. 33. Sectors contributing to GDP and energy-use sectors are different. "Commerce" includes wholesale and retail trade, finance, real estate, and business community, social, and personal services. "Other" includes mining, quarrying, utilities (electricity, gas, and water), construction, and nonprofit services to households.

(in constant yen) increased about six times, or almost 10 percent each year. Nearly half of the increase in manufacturing output was attributable to a twelvefold expansion of machinery production in the period 1960–79. Other branches of manufacturing increased their output by an impressive, but less phenomenal, 364 percent.[2]

Rapid economic growth was associated with an equally rapid increase in energy consumption (table 2-1). The ratio of energy use to GDP—a general indicator of energy intensity—in 1979 was not far from what it had been in 1960, although it rose to higher values between those years. Energy consumption per capita in 1979 was more than three and a half times what it was in 1960. The growth in GDP and energy consumption was not, however, constant from year to year, nor did these two variables always change at similar rates.

As table 2-1 indicates, the rates of growth of both GDP and energy consumption slackened in 1962, 1965, and 1971. But in all of these years the percentage increase in energy consumption did not fall as low as that of GDP. In 1974, for the first time since the end of World War II, Japan experienced an actual decline in GDP. On this occasion, energy consumption fell more than GDP; it decreased even more sharply in 1975, when GDP registered a small increase.

The differences in the behavior of GDP and energy consumption are not surprising. The failure of energy consumption to respond fully to declines in GDP growth rates in 1962, 1965, and 1971 reflects the lack of a close link between some energy requirements, such as space heating, and the level of economic activity. In 1974–75, the relatively great fall in energy consumption was the combined result of an economic slowdown and a quadrupling of oil prices.

The year 1973 was also a turning point in the intensity of energy use in Japan. From 1960 to 1973, the trend in the ratio of gross energy consumption to GDP was clearly upward. In 1974, the ratio turned down, and by 1979 it was 82 percent of its 1973 level.

The Arab-Israeli war of October 1973, the Arab oil embargo of 1973–74, and the huge increase in oil prices imposed by the Organization of Petroleum Exporting Countries (OPEC) had a profound effect on both

2. Japanese Ministry of International Trade and Industry, Research and Statistics Department, *Census of Manufactures, Report by Industries* (Tokyo: MITI, various years). Current yen were converted to 1970 yen by using the wholesale price index in Prime Minister's Office, Statistics Bureau, *Japan Statistical Yearbook* (Tokyo: Statistics Bureau, various years).

Table 2-1. *Population, Gross Domestic Product, and Energy Consumption in Japan, 1960–79*

Year	Population (millions)	Population growth rate (percent)	GDP (trillions of 1970 yen)	GDP growth rate (percent)	Final energy consumption (10 trillions of kilocalories)[a]	Energy consumption growth rate (percent)	GDP/population (thousands of yen per capita)	Energy/population (millions of kilocalories per capita)	Energy/GDP (kilocalories per yen)
1960	93.4	...	25.8	...	60.6	...	274.1	6.49	23.68
1961	94.3	0.93	29.3	13.56	69.6	14.85	311.1	7.39	23.74
1962	95.2	0.95	31.4	7.16	76.4	9.77	329.9	8.03	24.34
1963	96.2	1.02	34.7	10.51	85.6	12.03	360.9	8.91	24.68
1964	97.2	1.07	39.3	13.21	96.4	12.62	404.3	9.92	24.55
1965	98.3	1.13	41.3	5.12	106.6	10.54	420.2	10.85	25.82
1966	99.0	0.77	45.8	10.76	117.7	10.44	462.6	11.89	25.74
1967	100.2	1.17	51.5	12.56	136.4	15.86	513.8	13.61	26.50
1968	101.3	1.13	58.8	14.07	152.4	11.71	579.5	15.04	25.95
1969	102.5	1.19	65.8	12.17	176.6	15.90	642.4	17.22	26.81
1970	103.7	1.15	73.6	11.72	204.0	15.48	709.5	19.66	27.71
1971	105.1	1.35	77.4	5.11	218.4	7.09	736.6	20.80	28.24
1972	107.6	2.38	84.6	9.34	228.2	4.50	786.2	21.26	26.99
1973	109.1	1.39	93.0	9.95	260.0	13.90	852.4	23.91	27.96
1974	110.6	1.35	92.7	-0.34	256.5	-1.35	838.1	23.30	27.67
1975	111.9	1.24	93.9	1.05	243.7	-4.99	839.2	21.77	25.94
1976	113.1	1.03	100.0	6.80	253.9	4.20	884.4	22.45	25.38
1977	114.2	0.94	105.4	5.39	256.6	1.06	923.4	22.48	24.34
1978	115.2	0.89	111.6	5.87	262.7	2.40	969.0	22.81	23.54
1979	116.1	0.83	117.7	5.47	270.1	2.81	1,013.8	23.26	22.95
				Average annual growth rates					
1960–73	...	1.17	...	10.40		11.90	...	...	...
1973–79	...	1.10	...	4.04		0.69	...	...	...
1960–79	...	1.15	...	8.39		8.36	...	...	...

Sources: Population: Prime Minister's Office, Statistics Bureau, *Japan Statistical Yearbook, 1982* (Tokyo: Statistics Bureau, 1982), pp. 14–15; GDP: Organization for Economic Cooperation and Development, personal communication to the author for 1960–69 data; for 1970 on, see OECD, *National Accounts of OECD Countries, 1961–78* (Paris: OECD, 1980), p. 33; 1979 GDP estimated from growth rate of gross national product (GNP) reported in OECD, *OECD Economic Outlook*, no. 28 (Paris: OECD, December 1980), p. 81; energy consumption: Institute of Energy Economics (Japan) (IEE), *Energy Balance Tables* (Tokyo: IEE, various years). Numbers are rounded.

a. Final energy consumption is energy use by final consumers and does not include losses in the transformation sector. Energy consumption figures for 1960–64 are for fiscal years, which begin on April 1; figures for 1965–79 are for calendar years. Ten trillion kilocalories approximately equal a million metric tons of oil equivalent (t.o.e.).

economic growth and energy consumption. From 1960 to 1973, GDP grew at an average annual rate of 10.5 percent, and energy consumption grew at an average annual rate of 11.8 percent. From 1973 to 1979, GDP increased about 4.0 percent a year, and energy consumption increased about 0.6 percent.

Because of the marked difference in economic performance—and in energy consumption—before and after the 1973 oil crisis, energy developments in 1960–73 and 1974–79 will be analyzed separately. This approach makes it easier to explore the way in which Japan has adjusted to the higher oil prices imposed by OPEC in 1973–74, a matter taken up at length in chapter 4. The second oil crisis of 1979–80 will also be treated briefly in chapter 4, but for most purposes the analysis of past energy developments in Japan will end with 1979. The large increase in oil prices in 1980 was a major discontinuity that initiated a new sequence of events whose outcomes have not yet been fully revealed. With a few exceptions, 1980 data will be left to later chapters, where they provide a baseline for considering possible future developments.

Developments before the 1973–74 Oil Crisis

Simplified energy balances are given for fiscal year 1960 (table 2-2) and calendar year 1973 (table 2-3).[3] To facilitate structural comparison, these energy balances have been converted to percentages in tables 2-4 and 2-5. As these tables show, the 1960s and early 1970s brought dramatic changes in Japan's energy economy. Total use of energy (including transformation losses) more than quadrupled; the average annual rate of increase was 11.5 percent. Domestic energy production was unable to keep pace with the rapid growth in consumption. In 1960 Japan imported a little more than four-tenths of its energy. In 1973, only thirteen years later, nine-tenths of its energy was imported.

These figures are not substantially altered when account is taken of the energy embodied in Japan's nonfuel international trade.[4] Japan's

3. An energy balance for calendar year 1960 is not available. The Japanese fiscal year begins on April 1 of the year in question.

4. The energy embodied in Japan's international trade was estimated by using coefficients derived from 1972 input-output tables for the United States. See Bruce Hannon, Robert A. Herendeen, and Thomas Blazeck, "Energy and Labor Intensities for 1972," ERG Doc. 307 (University of Illinois at Urbana-Champaign, Energy Research Group, April 1981).

Table 2-2. *Energy Balance in Japan, Fiscal Year 1960*
Ten billions of kilocalories[a]

Item	Coal[b]	Oil[c]	Gas[d]	Hydro-electric[e]	Other[f]	Total elec-tricity	Total
Primary energy supply							
Domestic production	30,253	557	716	14,461	3,625	0	49,612
Net imports	5,998	29,623	0	0	0	0	35,621
Stock changes	1,099	−848	−2	0	0	0	249
Total	37,350	29,332	714	14,461	3,625	0	85,482
Transformation sector[g]	−11,016	−8,539	1,345	−14,461	0	7,820	−24,851
Final energy consumption, by sector							
Industrial	19,185	10,348	822	0	0	5,987	36,342
Residential and com- mercial	3,666	1,711	1,239	0	3,625	1,415	11,656
Transportation	3,483	7,520	0	0	0	418	11,421
Nonenergy uses of fuels[h]	0	1,214	0	0	0	0	1,214
Total	26,334	20,793	2,061	0	3,625	7,820	60,633

Source: Based on more detailed energy balance prepared by the IEE, *Energy Balance Tables*.

a. Ten billion kilocalories approximately equal a thousand t.o.e. Lines and columns may not add because of rounding.

b. Includes coking coal, steam coal, anthracite, lignite, coke oven gas, blast furnace gas, and briquets.

c. Includes crude oil and petroleum products.

d. Includes natural gas, liquefied natural gas, natural gas from coal mines, and town gas (gas manufactured from oil and coal derivatives).

e. Electricity generated by hydropower (and when applicable, by nuclear power) is given the same calorific value as a similar amount of electricity generated by thermal means.

f. Includes wood, charcoal, and nonhydroelectric autogeneration (electricity generated by the consumer).

g. The transformation sector records net inputs of primary energy used in generating electricity, refining oil, making coke and briquets from coal, and making town gas from oil and coal derivatives. The entry in the "gas" column for the transformation sector is positive because oil and coal, rather than gas, are used in making town gas.

h. In Japan, exclusively oil.

nonfuel exports contained somewhat more energy than did its nonfuel imports. The 1965 export surplus of embodied energy was 2.7 percent of total primary energy supply and 4.2 percent of net fuel imports. The 1973 surplus was 3.1 percent of primary energy supply and 3.4 percent of net fuel imports.

In 1960 domestic coal and imported oil were of approximately equal importance. Both provided about 35 percent of Japan's primary energy supply. By 1973 the share of imported oil had risen to nearly 78 percent, and the share of domestic coal had dropped to only about 4 percent. Over three-fourths of the oil imported in 1973 was from the Middle East. Physical limitations of Japan's coal resources would have precluded a rapid expansion of domestic coal production. In any event, Japan's coal industry could not compete with the cheap, plentiful, and convenient supplies of imported oil, and domestic coal output in 1973 was only about

Table 2-3. *Energy Balance in Japan, Calendar Year 1973*
Ten billions of kilocalories[a]

Item	Coal[b]	Oil[c]	Gas[d]	Nuclear[e]	Hydro-electric[e]	Other[f]	Total electricity	Total
Primary energy supply								
Domestic production	14,492	760	2,804	2,327	18,719	750	0	39,852
Net imports	40,739	271,548	2,569	0	0	0	0	314,856
Stock changes	1,232	−5,980	−50	0	0	0	0	−4,798
Total	56,463	266,328	5,323	2,327	18,719	750	0	349,910
Transformation sector[g]	−15,739	−90,562	2,244	−2,327	−18,719	0	35,166	−89,937
Final energy consumption, by sector								
Industrial	39,297	94,225	2,272	0	0	0	23,963	159,757
Residential and commercial	1,172	29,390	5,294	0	0	750	10,077	46,683
Transportation	255	44,390	1	0	0	0	1,126	45,772
Nonenergy uses of fuels[h]	0	7,761	0	0	0	0	0	7,761
Total	40,724	175,766	7,567	0	0	750	35,166	259,973

Source: Based on more detailed energy balance prepared by the IEE, *Energy Balance Tables*. Numbers are rounded.
a–h. See notes to table 2-2.

half of what it had been in 1960. Despite a large increase in coal imports, coal—both domestic and imported—accounted for only a sixth of total primary energy supply in 1973.

The major cause of the large increase in energy use from 1960 to 1973 was the rapid growth of the economy. Energy use increased more rapidly, however, than did the GDP. If the energy intensity of the Japanese economy (measured by the ratio of energy use to GDP) in 1973 had been as low as it was in 1960, the 1973 GDP could have been produced with 11 percent less energy. The rise in energy intensity would have been even greater had there not been a substantial increase in the thermal efficiency of energy use, an increase largely attributable to the shift from coal to oil (see the appendix).

The increase in energy intensity in Japan can be explained in large part by the 26.6 percent fall in real energy prices from 1960 to 1973.[5] If

5. J. Dunkerley, J. Alterman, and J. J. Schanz, Jr., *Trends in Energy Use in Industrial Societies*, EPRI EA-1471 (Palo Alto, Calif.: Electric Power Research Institute, 1980), p. 4-7.

Table 2-4. *Structure of Energy Supply and Disposition in Japan,*
Fiscal Year 1960
Percent[a]

Item	Coal[b]	Oil[c]	Gas[d]	Hydro-electric[e]	Other[f]	Total elec-tricity	Total
Primary energy supply							
Domestic production	35.4	0.7	0.8	16.9	4.2	. . .	58.0
Net imports	7.0	34.7	0.0	0.0	0.0	. . .	41.7
Stock changes	1.3	−1.0	*	0.0	0.0	. . .	0.3
Total	43.7	34.3	0.8	16.9	4.2	. . .	100.0
Disposition of energy supply,							
by sector							
Transformation sector[g]	12.9	10.0	−1.6	16.9	0.0	−9.1	29.1
Industrial	22.4	12.1	1.0	0.0	0.0	7.0	42.5
Residential and commercial	4.3	2.0	1.4	0.0	4.2	1.7	13.6
Transportation	4.1	8.8	0.0	0.0	0.0	0.5	13.4
Nonenergy uses of fuels[h]	0.0	1.4	0.0	0.0	0.0	0.0	1.4
Total	43.7	34.3	0.8	16.9	4.2	0.0	100.0

Source: Table 2-2.
* Less than 0.05 percent.
a. Lines and columns may not add because of rounding.
b–f,h. See notes to table 2-2.
g. In table 2-2 inputs to the transformation sector were treated as subtractions from primary energy supply that reduced the amount of energy available to the consuming sectors. Here the transformation sector is itself treated as a consuming sector.

the price elasticity of energy demand were as great as −0.5, the fall in prices would account for all of the increase in intensity. That is, if real energy prices had remained unchanged, the ratio of energy use to GDP would have been the same in 1973 as it was in 1960. In that event, the income elasticity of energy demand would have been 1.0, which happens to be its average level during this period.[6]

General explanations such as this, however, are of limited value. One also must explore developments in the four principal energy-consuming sectors: industrial, residential and commercial, transportation, and transformation.[7] One purpose of the examination of sectoral developments is to determine the relative influence of intensity and structural factors in changes in sectoral ratios of energy use to GDP (see the appendix).

6. OECD, International Energy Agency (IEA), *Energy Conservation in the International Energy Agency, 1978 Review* (Paris: OECD, 1979), p. 13.
7. Information on energy consumption in components of these sectors was obtained from Institute of Energy Economics (Japan), *Energy Balance Tables* (Tokyo: IEE, October 20, 1980). A fifth sector, nonenergy uses of fuels, accounted for only 1.4 percent of total energy disposition in 1973.

Table 2-5. *Structure of Energy Supply and Disposition in Japan, Calendar Year 1973*
Percent[a]

Item	Coal[b]	Oil[c]	Gas[d]	Nuclear[e]	Hydro-electric[e]	Other[f]	Total electricity	Total
Primary energy supply								
Domestic production	4.1	0.2	0.8	0.7	5.3	0.2	...	11.4
Net imports	11.6	77.6	0.7	0.0	0.0	0.0	...	90.0
Stock changes	0.4	−1.7	*	0.0	0.0	0.0	...	−1.4
Total	16.1	76.1	1.5	0.7	5.3	0.2	...	100.0
Disposition of energy supply, by sector								
Transformation[g]	4.5	25.9	−0.6	0.7	5.3	0.0	−10.1	25.7
Industrial	11.2	26.9	0.6	0.0	0.0	0.0	6.8	45.7
Residential and commercial	0.3	8.4	1.5	0.0	0.0	0.2	2.9	13.3
Transportation	0.1	12.7	0.0	0.0	0.0	0.0	0.3	13.1
Nonenergy uses of fuels[h]	0.0	2.2	0.0	0.0	0.0	0.0	0.0	2.2
Total	16.1	76.1	1.5	0.7	5.3	0.2	0.0	100.0

Source: Table 2-3.
* Less than 0.05 percent.
a. Lines and columns may not add because of rounding.
b–f, h. See notes to table 2-2.
g. See note g in table 2-4.

Energy use in the industrial sector (manufacturing, agriculture, forestry, fishing, and mining) grew 340 percent from 1960 to 1973, and the industrial sector's share in total energy disposition increased from 42.5 percent to 45.7 percent. Most of the energy used by industry was consumed in manufacturing (93 percent in 1960 and 94 percent in 1973). The ratio of energy use to GDP in manufacturing and in the industrial sector as a whole increased substantially during the 1960–73 period. Analysis of the manufacturing component of the industrial sector indicates that the increase in this ratio was not caused by a rise in the intensity of energy use, as might have been supposed. Energy intensity (measured by the ratio of energy use to value added) actually declined. The rise in the ratio of energy use to GDP was caused by the increased importance of manufacturing in the total economy, an increase large enough to overwhelm the effect of the fall in energy intensity.

The residential and commercial sector presents a somewhat different picture. Energy consumption in this sector grew slightly less rapidly from 1960 to 1973 than did total energy use. The increase in residential and commercial energy consumption can be explained as the combined effect of rising per capita incomes and falling energy prices (both expressed in real terms).

The ratio of energy use to GDP in this sector increased moderately from 1960 to 1973. In the residential component of the sector, however, the ratio declined. This decline can be explained by a fall in both the intensity of residential energy use (measured by the ratio of energy use to disposable income) and a structural factor (the ratio of disposable income to GDP). The change in intensity was twice as important as the structural change. The lag in residential energy consumption behind disposable income can be explained at least partly by a shift from use of coal, wood, and charcoal to use of oil, which is more efficient.

Although residential energy consumption grew less rapidly than did total energy use, the annual rate of increase (9.3 percent) was quite impressive. This growth reflects both an increase in space-heating requirements and the adoption of a variety of energy-using appliances by Japanese households.

Energy consumption in the transportation sector also lagged slightly behind total energy use in the 1960–73 period. The replacement of coal by oil in transportation uses was virtually completed during this period. The share of highway transportation in sectoral energy use increased greatly, and the share of railways fell. Total freight transportation lost ground to total passenger transportation.

The ratio of energy use to GDP increased both in passenger transportation and in the sector as a whole. The increase in this ratio in passenger transportation was entirely due to a shift to an overwhelming reliance on automobiles. The effect of this shift was greater than the combined effect of declines in both the intensity of energy use and the importance of passenger transportation in the total economy.

Transformation losses grew much less rapidly than did total energy use. In 1960 these losses were 29.1 percent of total primary energy supply; in 1973 they were 25.8 percent. In both 1960 and 1973 the generation of electricity accounted for about 80 percent of total transformation losses. The fall in the proportion of total energy supply lost in transformation processes must be largely credited to an increase in the thermal efficiency of electricity generation.

Government played a largely indirect role in the changes in energy supply and disposition before the 1973–74 oil crisis. Energy policy in this period emphasized diversification in fuels, the geographic origins of supplies, and the nationalities of energy-importing firms. This policy

Table 2-6. *Energy Balance in Japan, Calendar Year 1979*
Ten billions of kilocalories[a]

Item	Coal[b]	Oil[c]	Gas[d]	Nuclear[e]	Hydro-electric[e]	Other[f]	Total electricity	Total
Primary energy supply								
Domestic production	11,178	518	2,589	14,940	20,616	312	0	50,153
Net imports	40,154	277,740	18,523	0	0	0	0	336,417
Stock changes	466	−5,719	1	0	0	0	0	−5,252
Total	51,798	272,539	21,113	14,940	20,616	312	0	381,318
Transformation sector[g]	−14,467	−93,222	−11,670	−14,940	−20,616	−88	43,806	−111,197
Final energy consumption, by sector								
Industrial	35,973	85,228	2,020	0	0	0	26,941	150,162
Residential and commercial	1,358	30,543	7,423	0	0	224	15,567	55,115
Transportation	0	56,466	0	0	0	0	1,298	57,764
Nonenergy uses of fuels[h]	0	7,080	0	0	0	0	0	7,080
Total	37,331	179,317	9,443	0	0	224	43,806	270,121

Source: Based on more detailed energy balance prepared by the IEE, *Energy Balance Tables*.
a–h. See notes to table 2-2.

produced only limited results. The developmental policies of the government, however, indirectly contributed to increased energy consumption by stimulating rapid economic growth and by favoring the expansion of the relatively energy-intensive modern manufacturing sector.

Developments since the 1973–74 Oil Crisis

The oil crisis of 1973–74, particularly the large increase in the international price of crude oil, caused substantial changes in Japan's economy and in its use of energy. Seven years later, rates of economic growth before the crisis had not even been approached, and—contrary to the pattern before the crisis—total energy use had increased much less rapidly than the GDP. (See table 2-1 and the simplified energy balances for 1979 in tables 2-6 and 2-7.)

Japan's primary energy supply (and total energy use) was only 9 percent larger in 1979 than it had been in 1973. Energy consumption per capita fell in 1974 and 1975 and had barely regained the 1973 level by 1979. The fall of energy consumption per capita in 1974 and 1975 was the combined result of rising energy prices and declining GDP per capita.

The Energy Balance in Northeast Asia

Table 2-7. *Structure of Energy Supply and Disposition in Japan,*
Calendar Year 1979
Percent[a]

Item	Coal[b]	Oil[c]	Gas[d]	Nuclear[e]	Hydro-electric[e]	Other[f]	Total electricity	Total
Primary energy supply								
Domestic production	2.9	0.1	0.7	3.9	5.4	0.1	0.0	13.2
Net imports	10.5	72.8	4.9	0.0	0.0	0.0	0.0	88.2
Stock changes	0.1	− 1.5	*	0.0	0.0	0.0	0.0	− 1.4
Total	13.6	71.5	5.5	3.9	5.4	0.1	0.0	100.0
Disposition of energy supply, by sector								
Transformation[g]	3.8	24.4	3.1	3.9	5.4	0.1	− 11.5	29.2
Industrial	9.4	22.4	0.5	0.0	0.0	0.0	7.1	39.4
Residential and commercial	0.4	8.0	1.9	0.0	0.0	0.0	4.1	14.5
Transportation	0.0	14.8	0.0	0.0	0.0	0.0	0.3	15.1
Nonenergy uses of fuels[h]	0.0	1.9	0.0	0.0	0.0	0.0	0.0	1.9
Total	13.6	71.5	5.5	3.9	5.4	0.1	0.0	100.0

Source: Table 2-6.
* Less than 0.05 percent.
a. Lines and columns may not add because of rounding.
b–f, h. See notes to table 2-2.
g. See note to table 2-4.

The recovery of GDP per capita in 1976 was sufficiently strong to overcome the depressing effect that higher energy prices continued to have on consumption.

Japan's dependence on imported energy was slightly lower in 1979 (88.2 percent) than it had been in 1973 (90.0 percent). Because of a steep rise in production of nuclear energy, which Japan counts as domestic, domestic energy production increased by over 25 percent. Imported energy rose by about 7 percent, largely because of a huge increase in liquefied natural gas imports. Dependence on Arab countries for crude oil rose from 50.8 percent in 1974 to 62.9 percent in 1979.[8]

If the energy embodied in nonfuel international trade is considered, Japan's net dependence on imported energy is somewhat lower. Japan had an export surplus of embodied energy that was equivalent to 10.8 million metric tons of oil in 1973 and 75.2 million metric tons in 1979.

8. Percentages for 1979 were calculated from data in U.S. Central Intelligence Agency, *International Energy Statistical Review,* September 30, 1980, p. 6. Those for 1974 were calculated from data in U.S. Central Intelligence Agency, *International Energy Biweekly Statistical Review,* October 19, 1977, p. 7.

The 1979 surplus was over 20 percent of net energy imports; the surplus was 3.4 percent in 1973.

Significant changes occurred from 1973 to 1979 in the structure of Japan's primary energy supply. The relative importance of both oil and coal declined, and the shares of natural gas and nuclear energy increased.

The sectoral disposition of Japan's primary energy supply also changed significantly between 1973 and 1979, as the following percentage distribution by energy-consuming sector shows:

	1973	1979
Transformation	25.7	29.2
Industrial	45.7	39.4
Residential and commercial	13.3	14.5
Transportation	13.1	15.1
Nonenergy uses of fuels	2.2	1.9

By far the most important change in energy disposition was the decline in the share of the industrial sector. The shares of all other sectors except nonenergy uses of fuels (in Japan, exclusively oil) increased in both absolute and relative terms.

If the ratio of energy use to GDP had been as high in 1979 as it was in 1973, Japan would have needed a sixth more energy to produce its 1979 GDP. This marked fall in the overall energy intensity of the Japanese economy took place despite an increase in transformation losses (associated with a rise in the percentage of total primary energy supply consumed in the form of electricity) and a decline in the thermal efficiency of energy use (largely caused by a percentage shift in oil consumption away from the industrial sector and into the transportation sector, where oil is used less efficiently).

Most of the fall in the ratio of energy use to GDP after the 1973–74 oil crisis took place in the industrial sector. In only six years, from 1973 to 1979, the ratio of energy use to GDP in manufacturing industry dropped below the 1960 level. This decline was caused almost entirely by a decrease in the energy intensity of the sector, particularly of the energy-intensive subsector, and by a decrease in the share of energy-intensive industries in all manufacturing. These factors were sufficient to overcome an increase in the overall economic importance of the manufacturing sector.

An examination of four energy-intensive industries—iron and steel, petrochemicals, cement, and aluminum—shows that in all of them conservation efforts reduced energy consumption below what it would otherwise have been.[9] In the iron and steel industry, however, other factors (idle capacity, reduced use of scrap, antipollution controls, and the like) more than canceled conservation gains, and the energy consumed to produce a metric ton of steel rose 10.6 percent from 1973 to 1977. Two-thirds of conservation savings in the petrochemical industry were lost because of environmental regulations and changes in production processes.

Energy consumption in the residential and commercial sector increased about twice as rapidly from 1973 to 1979 as did total primary energy supply. About 80 percent of this increase can be attributed to a rise in GDP per capita; the remainder may be the result of a fall in real energy prices.

Residential energy consumption increased somewhat more rapidly from 1973 to 1979 than did consumption in the sector as a whole. The rate of increase, however, was only about 40 percent of the rate of increase from 1960 to 1973. The slowing of residential energy consumption was the result of a variety of factors, including a fall in energy use for cooking and space heating, a decrease in the rate of accumulation of household appliances, and an improvement in the efficiency of some appliances.

Energy consumption in the transportation sector increased from 1973 to 1979 about three times as fast as did total primary energy supply. Energy consumption continued to grow more rapidly in passenger transportation than it did in freight transportation. The share of railways in the sector's energy consumption continued to decline. In 1979 highway vehicles consumed 80 percent of the energy used by all forms of transportation. The primary driving force behind the rise of energy consumption in transportation was a rapid increase in the automobile inventory.

Transformation losses increased by nearly a quarter from 1973 to 1979. In both years, over three-fourths of the losses were in electricity generation, and over an eighth in petroleum refining. The thermal efficiency of electric power plants did not change appreciably. The net energy output of refineries, however, fell from 94.8 percent of the crude oil input in 1973 to 93.7 percent in 1979.

9. See IEE, "Present State and Future Potentials of Energy Conservation in Japan (Summary)," supplement to *Energy in Japan,* no. 45 (June 1979).

The 1973–74 oil crisis stimulated the Japanese government to be more active in the field of energy than it had in the period 1960–73. Efforts to diversify sources of energy were intensified. Oil stocks were built up, a variety of energy conservation measures were applied, and modest amounts were spent on energy research and development. Through diplomacy, economic aid, and investments, efforts were made to improve relations with energy-exporting countries.

The totality of government activity in energy is impressive, but Japanese energy policy since the oil crisis has been deliberate and gradual. There have been no crash programs, and—consistent with Japanese administrative style in other areas—Japanese energy policy has rarely been prescriptive. The government sets standards and provides guidance; by and large, the public complies.

As was true in the period before the 1973–74 oil crisis, the general economic policies of the Japanese government during and since the crisis have had an important effect on energy developments. Since the crisis, the government has been more concerned with stability than with rapid growth. As a result, the government's economic policies have contributed to the marked slowing down in the rate of increase in energy consumption. They have, therefore, complemented the government's energy conservation efforts.

Comparison of the Experience of Japan and Three Other Industrial Countries

A fuller understanding of energy developments in Japan during the 1960s and 1970s may be obtained by comparing the experience of Japan with that of three other non-Communist industrialized countries—the United States, France, and West Germany. Among these four countries, Japan ranks second to the United States in GDP, total energy consumption, and the overall energy intensity of its economy. Energy developments in Japan in 1960–80 have been more comparable, however, with developments in France and West Germany (table 2-8).

During this period coal declined in importance in Japan, France, and West Germany, and was replaced by oil and other fuels. In the United States, the shift away from coal was completed before 1960. Changes in the disposition of energy supplies were in general also similar in Japan, France, and West Germany. In these three countries, the importance of

Table 2-8. *Population, GDP, and Energy Supply of Japan and Three Other Industrialized Countries, 1960, 1973, and 1980*

Item	Japan	United States	France	West Germany
Population (millions)				
1960	93.3	180.7	45.7	55.4
1973	108.7	211.9	52.1	62.0
1980	116.8	227.7	53.7	61.6
GDP (billions of 1975 dollars)				
1960	136.8	932.4	162.0	239.7
1973	492.5	1,561.9	327.6	425.9
1980	639.4	1,832.2	398.6	500.4
Primary energy supply (millions of metric tons of oil equivalent)				
1960	94.7	1,014.2	86.8	145.3
1973	337.8	1,756.3	183.7	265.8
1980	372.7	1,809.5	198.2	272.2
GDP per capita (thousands of 1975 dollars)				
1960	1.5	5.2	3.5	4.3
1973	4.5	7.4	6.3	6.9
1980	5.5	8.0	7.4	8.1
Energy use per capita (thousands of metric tons of oil equivalent)				
1960	1.0	5.6	1.9	2.6
1973	3.1	8.3	3.6	4.3
1980	3.2	7.9	3.7	4.4
Energy/GDP (thousands of kilocalories per 1975 dollar)				
1960	6.9	10.9	5.4	6.1
1973	6.9	11.2	5.7	6.3
1980	5.8	9.9	5.0	5.4

Sources: Population and GDP: 1960, OECD, *National Accounts of OECD Countries, 1950–78*, vol. 1, pp. 86, 89; 1973 and 1980, OECD, *National Accounts, 1951–80*, vol. 1, pp. 86, 89. Energy: 1960, OECD, International Energy Agency (IEA), *Energy Balances of OECD Countries, 1960–74*, pp. 101, 114, 116; 1973, OECD, IEA, *Energy Balances, 1973–75*, pp. 45, 48, 72, 75; 1980, OECD, IEA, *Energy Balances, 1976–80*, pp. 36, 41, 76, 81.

the transformation sector declined, and the importance of the transportation and "other" sectors increased.[10] In the United States, the share of the transformation sector in total energy disposition increased, and the share of the "other" sector declined. In 1980 Japan differed from the other three countries in two important respects: the share of its industrial sector in total energy disposition was much larger, and the share of its "other" sector was much smaller (tables 2-9 and 2-10).

The energy intensities of all four economies, measured by the ratio of

10. The "other" sector in OECD usage is equivalent to the "residential and commercial" sector in Japanese energy statistics.

Table 2-9. *Sources of Total Energy Supply of Japan and Three Other Industrialized Countries, Selected Years, 1960–80*

Percent

Country and year	Coal	Oil	Gas	Nuclear	Hydro-electric	Imported electricity	Total
Japan							
1960	52.39	31.62	0.86	0.0	15.14	0.0	100
1965	33.92	52.49	1.22	0.0	12.36	0.0	100
1970	22.67	68.70	1.32	0.40	6.91	0.0	100
1973	16.86	75.62	1.62	0.70	5.20	0.0	100
1979	14.77	70.05	5.07	4.57	5.54	0.0	100
1980	17.82	64.68	5.87	5.43	6.19	0.0	100
United States							
1960	22.71	44.62	28.98	0.01	3.64	0.04	100
1965	22.25	43.31	30.39	0.08	3.97	*	100
1970	19.21	43.73	32.73	0.36	3.95	0.01	100
1973	18.67	45.05	31.03	1.25	3.93	0.07	100
1979	20.48	46.30	25.77	3.55	3.81	0.09	100
1980	22.38	43.42	26.53	3.63	3.91	0.13	100
West Germany							
1960	75.53	21.53	0.51	0.0	2.19	0.25	100
1965	54.79	41.58	1.39	0.02	2.04	0.20	100
1970	37.67	54.05	5.52	0.63	1.84	0.28	100
1973	31.36	55.46	10.35	1.08	1.43	0.33	100
1979	27.88	50.29	16.58	3.64	1.59	0.02	100
1980	30.68	47.95	15.91	3.70	1.58	0.18	100
France							
1960	54.59	30.99	2.83	0.05	11.56	−0.01	100
1965	40.72	45.13	3.82	0.23	10.03	0.07	100
1970	25.63	58.58	5.56	0.93	9.31	−0.03	100
1973	16.71	67.61	7.48	1.95	6.39	−0.14	100
1979	17.18	59.69	9.99	3.94	8.99	0.19	100
1980	18.01	55.14	11.06	7.27	8.39	0.14	100

Source: Percentages calculated from quantitative data in OECD, IEA, *Energy Balances*, 1976, 1977, 1981, and 1982.
* Negligible.

total energy use to GDP, were lower in 1980 than in 1960. Japan experienced the greatest decline (15.8 percent), followed by West Germany (10.3 percent). Energy intensity fell 9.3 percent in the United States, 7.2 percent in France. It is difficult to relate these changes in energy intensity to gross structural economic changes such as an increase in the share of manufacturing in GDP. The effects of certain other

Table 2-10. *Disposition of Total Energy Supply of Japan*
and Three Other Industrialized Countries by Sector,
Selected Years, 1960–80
Percent

Country and sector	1960	1965	1970	1973	1979	1980
Japan						
Transformation	35.3	29.1	23.1	28.0	27.2	27.8
Industrial	38.6	35.7	37.9	38.8	38.6	38.6
Transportation	12.4	12.0	11.3	11.7	14.3	14.9
Other[a]	11.8	14.7	16.6	19.5	16.4	16.0
Nonenergy uses of fuels	1.9	8.6	11.1	2.0	3.5	2.7
Total	100.0	100.0	100.0	100.0	100.0	100.0
United States						
Transformation	20.2	21.8	22.2	25.3	27.7	28.5
Industrial	26.8	26.9	25.5	23.7	21.6	21.0
Transportation	23.1	22.6	22.2	23.3	24.6	24.1
Other[a]	26.2	25.4	26.0	24.7	22.5	22.8
Nonenergy uses of fuels	3.7	3.2	4.1	3.1	3.6	3.6
Total	100.0	100.0	100.0	100.0	100.0	100.0
West Germany						
Transformation	29.7	26.9	25.0	23.5	25.6	27.5
Industrial	33.4	30.7	28.6	31.1	26.8	27.0
Transportation	10.9	11.2	11.8	12.3	13.7	15.0
Other[a]	23.8	26.6	28.4	29.5	31.0	28.8
Nonenergy uses of fuels	2.2	4.6	6.2	3.6	3.0	1.7
Total	100.0	100.0	100.0	100.0	100.0	100.0
France						
Transformation	26.9	25.5	22.9	20.9	22.4	27.5
Industrial	33.5	33.0	33.1	29.7	30.1	27.6
Transportation	13.4	13.7	14.1	14.7	17.1	16.6
Other[a]	23.4	24.7	24.3	29.6	27.4	25.6
Nonenergy uses of fuels	2.7	3.1	5.5	5.0	2.9	2.6
Total	100.0	100.0	100.0	100.0	100.0	100.0

Source: Same as table 2-9.
a. Includes mining, quarrying, utilities (electricity, gas, and water), construction, and nonprofit services to households; OECD equivalent of "residential and commercial" sector in other tables.

economic changes (such as thermal efficiency, transformation losses, the energy content of nonfuel trade, and energy prices) on energy intensity can more easily be identified.

The shift from coal to other fuels caused the thermal efficiency of energy use to rise in Japan, France, and West Germany from 1960 to 1973. These countries therefore were able to consume more useful energy per unit of GDP without significantly increasing their intensity of energy use in gross terms. Thermal efficiency in the United States benefited from the shift from coal before 1960, and energy intensity in the United States increased somewhat in the period 1960–73. In all four countries, thermal efficiency changed very little from 1973 to 1980.

A decline in transformation losses over the period 1960–80 in Japan, France, and West Germany held down energy intensity. An increase in such losses in the United States had the opposite effect.

The nonfuel exports of all four countries contained more embodied energy than did their nonfuel imports. In all cases, the export surpluses of embodied energy were much larger in 1980 than in 1965. In Japan the export surplus of embodied energy in 1979 was 25.8 percent of total energy supply. In West Germany it was 33.9 percent. The export surpluses of embodied energy in France and the United States were less important—8.3 percent and 3.6 percent of total energy supplies, respectively. To the extent that these countries exported more embodied energy than they imported, they were required to use more energy to produce their GDPs.

Real energy prices to final consumers declined in all four countries from 1960 to 1973 and rose (although not continuously) from 1973 to 1980.[11] Japan experienced the largest price decrease in the first period and the largest increase in the second. Price changes encouraged increased energy intensity before the 1973–74 oil crisis and discouraged it afterward.

Changes in overall energy intensity are the net result of changes in the ratios of energy use to GDP of the energy-consuming sectors. The decline in energy intensity in Japan, France, and West Germany from 1960 to 1980 is largely a result of decreases in these ratios within the transformation and industrial sectors. The decline in energy intensity in the United States was mainly caused by changes in the industrial and "other" sectors.

11. "Prospects for Recovery," *OECD Observer,* no. 108 (January 1981), p. 17.

The decreases in the ratios of energy use to GDP within the transformation sectors of Japan, France, and West Germany from 1960 to 1980 were associated with substantial increases in the efficiency of electricity generation. The greatest increase was in Japan. Generation efficiency in the United States in 1980 was approximately what it had been in 1960.

In all four countries, the fall in the ratio of energy use to GDP in the industrial sector from 1960 to 1980 was attributable to a decrease in the amount of energy required to produce a unit of industrial value added. This intensity effect was strong enough in every case to overcome a structural effect—an increase in the importance of industry in the total economy. Lack of suitable data precludes comparable structure-intensity analyses of changes in the ratios of energy use to GDP of the entire transportation and "other" sectors.

An analysis of the highway transportation subsector, however, shows that structural change—an increase in vehicle-kilometers traveled per unit of GDP—was the main cause of an increase in the ratio of energy use to GDP in three of the four countries.[12] The exception was West Germany, where a rise in energy intensity (measured by energy consumption per vehicle-kilometer) was the principal cause of an increase in the ratio.

An analysis of changes in the ratio of energy use to GDP in the residential subsector (which accounts for the largest part of energy consumption in the "other" sector) produced mixed results. In the period before 1973, this ratio declined in Japan and the United States but increased in France and West Germany. In all four countries, a change in intensity (the amount of energy consumed per unit of disposable income) was the principal cause of change in the ratio of energy use to GDP. After 1973, the ratio changed very little in Japan and West Germany. Decreases in the other two countries were the combined result of lower energy intensity and structural change (a reduction in disposable income relative to GDP). In the United States, the intensity effect was far more important than the structural effect; in France, the structural effect was somewhat greater.[13]

In general, changes in the intensity of energy use appear to have been

12. Data on vehicle-kilometers were obtained from International Road Federation, *World Road Statistics* (Geneva, Switzerland: IRF, various issues).
13. GDP and disposable income figures are given in World Bank, *World Tables: The Second Edition (1980)* (Johns Hopkins University Press for the World Bank, 1980), and in OECD, *National Accounts* (various issues).

more important than structural changes in explaining increases or decreases in sectoral and subsectoral ratios of energy use to GDP. In West Germany, energy intensity was found to be the main cause in every case analyzed. The results in France were more evenly divided between the two effects. In Japan and the United States, change in energy intensity was the dominant cause of change in industrial and residential ratios of energy use to GDP, and structural change was the main explanation for change in the ratio within the highway transportation subsector.

Governments have influenced energy developments principally through policies designed to promote economic growth or stability. Before the 1973–74 oil crisis, none of the four governments had a comprehensive energy policy.[14] Explicit energy policies were directed toward narrower problems, such as ailing domestic coal industries (in all four countries), the threat of cheap foreign oil to the domestic oil industry (in the United States), the power of foreign oil companies (in Japan, France, and, to a lesser extent, West Germany), and excessive dependence on imported oil (mostly in Japan). Over the long run, the commitments made by the four governments to the development of nuclear energy may have a greater effect on national energy futures than any other explicit energy policies pursued before the 1973–74 oil crisis.

Since the oil crisis, all four countries have tried to cut down their dependence on oil by shifting to other fuels and by reducing the energy intensity of their economies. Between 1973 and 1980, France reduced the share of oil in its total energy supply by 18.4 percent. West Germany and Japan achieved reductions of 13.5 percent and 14.5 percent, respectively. The share of oil in the U.S. energy supply decreased 3.6 percent. Japan led the four in lessening the energy intensity of its economy, with

14. For a description of Japan's energy policies before the first oil crisis, see Joseph A. Yager and Eleanor B. Steinberg, *Energy and U.S. Foreign Policy* (Cambridge, Mass.: Ballinger, 1974), pp. 149–60. Sources consulted concerning French and West German energy policies in this period include Guy de Carmoy, *Energy for Europe: Economic and Political Implications* (Washington, D.C.: American Enterprise Institute for Public Policy Research, 1977); Horst Mendershausen, *Coping with the Oil Crisis: French and German Experiences* (Johns Hopkins University Press for Resources for the Future, 1976); and Philip H. Trezise, "Western Europe," in Yager and Steinberg, *Energy and U.S. Foreign Policy*. The principal source of information on U.S. energy policy during this period was Craufurd D. Goodwin, ed., *Energy Policy in Perspective: Today's Problems, Yesterday's Solutions* (Brookings Institution, 1981), especially the chapter on the Kennedy administration by William J. Barber and the chapter on the Johnson administration by James L. Cochrane.

a reduction of 15.0 percent from 1973 to 1980. West Germany achieved a reduction of 13.1 percent, and both the United States and France realized reductions of about 12 percent.

Decreases in energy intensity may in part have been caused by governmental programs to encourage energy conservation. The main credit, however, most probably goes to higher energy prices. The policies of the four governments with respect to energy prices have tended to converge: in all four countries, increases in the price of imported oil were passed onto final consumers more fully in the 1979–80 oil crisis than in the 1973–74 crisis.[15]

15. "Prospects for Recovery," p. 16.

CHAPTER THREE

Energy Developments in South Korea and Taiwan in the 1960s and 1970s

In the past two decades South Korea and Taiwan have achieved high rates of economic growth despite limited endowments of natural resources. South Korea may have more coal than Taiwan, but thus far it has found no natural gas or oil. Taiwan produces a significant amount of natural gas and a small amount of oil.[1] For both South Korea and Taiwan, economic growth has brought increased dependence on imported energy.

Developments in South Korea

South Korea's gross domestic product grew at an average annual rate of 8.8 percent from 1960 to 1979. Over the same period, population expanded at an average annual rate of 2.2 percent. The rate of increase in GDP per capita was therefore held to 6.5 percent annually (table 3-1).

---

1. South Korea's proven coal reserves are about 116 million metric tons. Probable reserves are 76–126 million metric tons, and possible reserves are estimated at 157–393 million metric tons. See Argonne National Laboratory, *Republic of Korea/United States Cooperative Energy Assessment* (Springfield, Va.: National Technical Information Service, 1981), p. 3-30. Taiwan's deposits of hydrocarbons are estimated to be 200 million metric tons of coal, 24 billion cubic meters of natural gas, and 1.6 million kiloliters of oil. See Ministry of Economic Affairs, Energy Committee, *The Energy Situation in Taiwan, Republic of China* (Taipei: MEA, 1982), p. 6.

Table 3-1. *Population, Gross Domestic Product, and Energy Consumption in South Korea, 1960–79*

Year	Population (thousands)	Population growth rate (percent)	GDP (billions of 1975 won)	GDP growth rate (percent)	Energy consumption (10 billions of kilocalories)[a]	Energy consumption growth rate (percent)[b]	GDP/ population (thousands of won per capita)	Energy/ population (millions of kilocalories per capita)	Energy/GDP (kilocalories per won)
1960	25,013	n.a.	2,663	n.a.	n.a.	n.a.	106.46	n.a.	n.a.
1961	25,766	3.01	2,828	6.20	9,859	n.a.	109.76	3.83	34.86
1962	26,513	2.90	2,877	1.73	10,470	6.20	108.51	3.95	36.39
1963	27,262	2.83	3,139	9.11	11,060	5.64	115.14	4.06	35.23
1964	27,984	2.65	3,448	9.84	11,592	4.81	123.21	4.14	33.62
1965	28,705	2.58	3,630	5.28	12,123	4.58	126.46	4.22	33.40
1966	29,436	2.55	4,069	12.09	13,095	8.02	138.23	4.45	32.18
1967	30,131	2.36	4,282	5.23	13,890	6.07	142.11	4.61	32.44
1968	30,838	2.35	4,741	10.72	15,550	11.95	153.74	5.04	32.80
1969	31,544	2.29	5,386	13.60	17,395	11.86	170.75	5.51	32.30
1970	32,241	2.21	5,825	8.15	19,731	13.43	180.67	6.12	33.87
1971	32,883	1.99	6,372	9.39	21,266	7.78	193.78	6.47	33.37
1972	33,505	1.89	6,739	5.76	22,047	3.67	201.13	6.58	32.72
1973	34,103	1.78	7,729	14.69	25,282	14.67	226.64	7.41	32.71
1974	34,692	1.73	8,344	7.96	25,501	0.87	240.52	7.35	30.56
1975	35,281	1.70	9,004	7.91	27,067	6.14	255.21	7.67	30.06
1976	35,860	1.64	10,217	13.47	29,796	10.08	284.91	8.31	29.16
1977	36,436	1.61	11,207	9.69	33,067	10.98	307.58	9.08	29.51
1978	37,019	1.60	12,401	10.65	36,136	9.28	334.99	9.76	29.14
1979	37,605	1.58	13,264	6.96	40,475	12.01	352.72	10.76	30.51
Average annual growth rate									
1965–73	...	2.22	...	9.43	...	9.11	...	...	...
1973–79	...	1.66	...	9.44	...	8.23	...	...	...
1965–79	...	1.99	...	9.44	...	8.76	...	...	...

Sources: Population: Economic Planning Board (EPB), *Handbook of Korean Economy, 1979* (Seoul: EPB, 1979), pp. 340–41; entries for 1960 estimated by using rate of increase; GDP: Bank of Korea, *Economic Statistics Yearbook, 1980* (Seoul: BOK, 1980), pp. 272–73, and EPB, *Korea Statistical Yearbook, 1980* (Seoul: EPB, 1980), p. 469; energy: based on Ho Tak Kim, ''Korea's Energy Experience in the '70s and Short- to Medium-Term Options Open to Korea'' (Seoul: Korea Energy Research Institute, September 1980, draft), appendix table 4, and Korea Energy Research Institute, *Basic Data for Energy Policy Study* (Seoul: KERI, June 1980: in Korean), p. 1.

n.a. Not available.

a. Energy consumption figures are total energy use (equal to total primary energy supply). Electricity consumption adjusted by the ratio of 860 kilocalories per kilowatt hour. Ten billion kilocalories approximately equal a thousand metric tons of oil equivalent (t.o.e.).

b. Average growth rates for energy consumption cover the period 1963–76.

Rapid growth brought major changes in the structure of Korea's economy, as is shown by a comparison of percentage contributions in the sectoral breakdown of GDP for 1960 and 1979:[2]

	1960	1979
Agriculture	47.2	21.2
Manufacturing	7.2	29.1
Commerce	29.8	29.0
Transportation	2.0	8.1
Government services	9.4	3.2
Other	4.4	9.4

The decline in the share of agriculture and the increase in the shares of manufacturing and transportation are a pattern commonly seen in a developing country. In 1979, however, Korea still had a sizable agricultural sector.

From 1961 to 1979 GDP grew more rapidly than total energy use. In 1979 about 12.5 percent less energy was needed to produce a unit of GDP than in 1961 (table 3-1). The 1973–74 oil crisis had a greater immediate impact on energy use than it did on GDP. Energy use and GDP both grew rapidly in 1973, each by about 14.7 percent. In 1974 energy use increased only 0.9 percent, but GDP grew 8.0 percent—far below the rate of increase in 1973, yet not much below the average of 8.8 percent for the period 1961–73. In the 1979 oil crisis, energy consumption continued to grow rapidly, whereas the rate of growth of GDP dropped substantially.

Energy balance tables for South Korea are presented for 1965, 1973, and 1979 (tables 3-2 through 3-7; information to prepare a table for 1960 is not available). As was done for Japan, the components of the energy balances have also been expressed as percentages of total primary energy supply.

In 1965 Korea's total primary energy supply, including noncommercial fuels, was the equivalent of about 12.1 million metric tons of oil. In 1979 total primary energy supply had increased to the equivalent of about 40.5 million metric tons of oil. Dependence on imported energy, mostly oil, grew from 11.7 percent of total supply in 1965 to 79.5 percent in 1979.

The structure of Korea's energy supply changed greatly from 1965 to

2. Bank of Korea, *Economic Statistics Yearbook, 1980* (Seoul: BOK, 1980), pp. 272–73; Economic Planning Board, *Korea Statistical Yearbook, 1981* (Seoul: EPB, 1981), p. 469. "Agriculture" includes forestry and fishing; "transportation" includes storage and communications.

Table 3-2. *Energy Balance in South Korea, 1965*
Ten billions of kilocalories[a]

Item	Coal[b]	Oil[c]	Hydro-electric[d]	Other[e]	Total elec-tricity[f]	Total
Primary energy supply						
Domestic production	5,278	0	178	5,142	0	10,598
Net imports	−63	1,476	0	0	0	1,413
Stock changes	153	−37	0	0	0	116
Total	5,368	1,439	178	5,142	0	12,127
Transformation and statistical difference[g]	−712	−131	−178	0	213	−808
Final energy consumption, by sector						
Industrial	1,299	688	0	667	127	2,781
Residential and commercial	2,454	33	0	4,391	55	6,933
Transportation	149	503	0	8	4	664
Public and other	754	84	0	·76	27	941
Total	4,656	1,308	0	5,142	213	11,319

Source: General Ho Tak Kim, *Korea's Energy Experience in the '70s,* appendix tables 1, 2, and 4; electricity data: EPB, *Major Statistics of Korean Economy, 1980* (Seoul: EPB, 1980), p. 97. The original data apparently came from the Korea Energy Research Institute, *Basic Data for Energy Policy Study,* pp. 10–12. The sectoral energy consumption figures for 1961 through 1976 were from a table compiled by the Korea Development Institute; the data for 1977 through 1979 were from another series, by the Ministry of Energy. The overlapping years, 1975 through 1977, showed identical total consumption figures in both series. The sectoral distribution, however, differed slightly, possibly because of a difference in sectoral definition.

a. Ten billion kilocalories approximately equal a thousand t.o.e.

b. Includes coal and coal products.

c. Includes crude oil and petroleum products.

d. Electricity generated by hydropower was given a hypothetical thermal input value on the assumption that generating efficiency was 35.1 percent; the conversion ratio was therefore 2,450 kilocalories per kilowatt hour.

e. Includes wood, charcoal, and crop residue.

f. Total electricity consumption was derived by converting data in kilowatt hours to kilocalories, at a rate of 860 kilocalories per kilowatt hour, and was allocated among consuming sectors in the same proportions as unadjusted figures in the first of the sources listed above. No electricity was generated by nuclear power plants until 1978.

g. Transformation losses in the production of coal and petroleum products are assumed to be included in sectoral consumption figures. Transformation losses in generating electricity are included here.

1979. In 1965 coal provided 44.3 percent of total primary energy supply, and noncommercial fuels provided 42.4 percent. These sources of energy steadily lost ground to oil, which provided 61.0 percent of total supply in 1979 compared with only 11.9 percent in 1965. In 1979 the shares of coal and noncommercial fuels had dropped to 28.5 percent and 7.1 percent, respectively. Hydropower was a minor source of energy throughout the period (1.4 percent of total supply in both 1965 and 1979). Nuclear power came on the scene only in 1978; its share of total supply in 1979 was 1.9 percent.

The disposition of Korea's primary energy supply also changed

Table 3-3. *Structure of Energy Supply and Disposition in South Korea, 1965*
Percent[a]

Item	Coal[b]	Oil[c]	Hydro-electric[d]	Other[e]	Total electricity[f]	Total
Primary energy supply						
Domestic production	43.5	0.0	1.5	42.4	0.0	87.4
Net imports	−0.5	12.2	0.0	0.0	0.0	11.7
Stock changes	1.3	−0.3	0.0	0.0	0.0	1.0
Total	44.3	11.9	1.5	42.4	0.0	100.0
Disposition of energy supply, by sector						
Transformation and statistical differences[g]	5.9	1.1	1.4	0.0	−1.8	6.6
Industrial	10.7	5.7	0.0	5.5	1.0	22.9
Residential and commercial	20.2	0.3	0.0	36.2	0.5	57.2
Transportation	1.2	4.1	0.0	0.1	*	5.5
Public and other	6.2	0.7	0.0	0.6	0.2	7.8
Total	44.3	11.9	1.4	42.4	0.0	100.0

Source: Table 3-2.
* Less than 0.05 percent.
a. Numbers are rounded.
b–g. See notes to table 3-2.

substantially from 1965 to 1979. The residential and commercial sector dropped from 57.2 percent of total supply in 1965 to 31.8 percent in 1979. Korea's rapid industrialization was reflected in the increase of the share of the industrial sector, from 22.9 percent in 1965 to 37.6 percent in 1979.

The energy intensity of the Korean economy declined from 33.4 kilocalories per won in 1965 to 30.5 kilocalories per won in 1979. This decline was largely caused by a sharp drop in the ratio of energy use to GDP in the residential and commercial sector. The increase in this ratio, a general indicator of energy intensity, in the industrial and transportation sectors, however, offset over half of the decline in the residential and commercial sector.

The thermal efficiency of energy use increased from 33.2 percent in 1965 to 52.9 percent in 1979. This rise is partly attributable to the increased importance of the industrial sector (which uses all fuels most efficiently) and to a shift from less efficient fuels (noncommercial fuels and coal) to more efficient ones (oil and electricity). The increase in thermal efficiency contributed significantly to the decrease in the energy

Table 3-4. *Energy Balance in South Korea, 1973*
Ten billions of kilocalories[a]

Item	Coal[b]	Oil[c]	Hydro-electric[d]	Other[e]	Total elec-tricity[f]	Total
Primary energy supply						
Domestic production	6,989	0	315	3,672	0	10,976
Net imports	290	13,313	0	. . .	0	13,603
Stock changes	392	311	0	. . .	0	703
Total	7,671	13,624	315	3,672	0	25,282
Transformation and statistical						
difference[g]	−411	−3,088	−315	. . .	1,063	−2,751
Final energy consumption,						
by sector						
Industrial	1,151	4,361	0	497	675	6,684
Residential and commercial	5,603	636	0	3,133	296	9,668
Transportation	38	3,452	0	. . .	8	3,498
Public and other	468	2,087	0	42	84	2,681
Total	7,260	10,536	0	3,672	1,063	22,531

Sources: Same as table 3-2.
a–g. See notes to table 3-2.

intensity (measured by gross energy consumption per unit of GDP) of the Korean economy. The ratio of useful energy consumption to GDP rose substantially from 1965 to 1979.

Transformation losses were 12.1 percent of total energy requirements in 1967 and 19.9 percent in 1977. They therefore caused the economy to be more energy intensive than it otherwise would have been. The energy content of Korea's nonfuel international trade had a small, opposite effect. In 1965 the energy embodied in Korea's nonfuel imports was the equivalent of about 1.3 million metric tons of oil greater than the energy embodied in its nonfuel exports. In 1979 the import surplus of embodied energy was the equivalent of approximately 4.9 million metric tons of oil.

The rapid increase in GDP was the main cause of the rise in energy consumption in the 1960s and 1970s. Rising real energy prices somewhat moderated the effect of continued rapid economic growth on energy consumption.

The Korean government has influenced energy developments principally through actions designed to promote economic development. Before 1973 the government does not appear to have had a general

Table 3-5. *Structure of Energy Supply and Disposition in South Korea, 1973*
Percent[a]

Item	Coal[b]	Oil[c]	Hydro-electric[d]	Other[e]	Total elec-tricity[f]	Total
Primary energy supply						
Domestic production	27.6	0.0	1.2	14.5	0.0	43.4
Net imports	1.1	52.7	0.0	0.0	0.0	53.8
Stock changes	1.6	1.2	0.0	0.0	0.0	2.8
Total	30.3	53.9	1.2	14.5	0.0	100.0
Disposition of energy supply, by sector						
Transformation and statistical difference[g]	1.6	12.2	1.2	0.0	−4.2	10.9
Industrial	4.6	17.2	0.0	2.0	2.7	26.4
Residential and commercial	22.2	2.5	0.0	12.4	1.2	38.2
Transportation	0.1	13.7	0.0	0.0	*	13.8
Public and other[h]	1.8	8.3	0.0	0.2	0.3	10.6
Total	30.3	53.9	1.2	14.5	0.0	100.0

Sources: Same as table 3-2; based on table 3-4.
* Less than 0.05 percent.
a. Numbers are rounded.
b–h. See notes to table 3-2.

Table 3-6. *Energy Balance in South Korea, 1979*
Ten billions of kilocalories[a]

Item	Coal[b]	Oil[c]	Nu-clear[d]	Hydro-electric[d]	Other[e]	Total elec-tricity[f]	Total
Primary energy supply							
Domestic production	9,377	0	772	570	2,892	0	13,611
Net imports	2,745	29,425	0	0	0	0	32,170
Stock changes	−571	−4,735	0	0	0	0	−5,306
Total	11,551	24,690	772	570	2,892	0	40,475
Transformation and statistical difference[g]	−352	−7,086	−772	−570	0	2,678	−6,102
Final energy consumption, by sector							
Industrial	3,777	9,439	0	0	132	1,883	15,231
Residential and commercial	7,343	2,236	0	0	2,728	584	12,891
Transportation	1	4,303	0	0	0	32	4,336
Public and other	78	1,626	0	0	32	179	1,915
Total	11,199	17,604	0	0	2,892	2,678	34,373

Sources: Same as table 3-2.
a–g. See notes to table 3-2.

Table 3-7. *Structure of Energy Supply and Disposition in South Korea, 1979*
Percent[a]

Item	Coal[b]	Oil[c]	Nuclear[d]	Hydro-electric[d]	Other[e]	Total electricity[f]	Total
Primary energy supply							
Domestic production	23.2	0.0	1.9	1.4	7.1	0.0	29.7
Net imports	6.8	72.7	0.0	0.0	0.0	0.0	79.5
Stock changes	−1.4	−11.7	0.0	0.0	0.0	0.0	−13.1
Total	28.5	61.0	1.9	1.4	7.1	0.0	100.0
Disposition of energy supply, by sector							
Transformation and statistical difference[g]	0.9	17.5	1.9	1.4	0.0	−6.6	15.1
Industrial	9.3	23.3	0.0	0.0	0.3	4.7	37.6
Residential and commercial	18.1	5.5	0.0	0.0	6.7	1.4	31.8
Transportation	*	10.6	0.0	0.0	0.0	0.1	10.7
Public and other	0.2	4.0	0.0	0.0	0.1	0.4	4.7
Total	28.5	61.0	1.9	1.4	7.1	0.0	100.0

Sources: Same as table 3-2; based on table 3-6.
* Less than 0.05 percent.
a. Numbers are rounded.
b–g. See notes to table 3-2.

energy policy. Since 1973 it has tried, with little success, to diversify the geographic sources of energy imports, but it has made progress in substituting coal and nuclear energy for oil. The government's price policies have been generally calculated to further energy conservation. The government also has adopted several other measures to reduce energy consumption, including a program to finance improvements in the efficiency of energy use by manufacturing industries.

Developments in Taiwan

In 1960 Taiwan was still a poor country with a GDP per capita of roughly $370 (in 1979 New Taiwan dollars converted at a rate of 40 to the dollar). The economy of Taiwan, however, was by no means stagnant. From 1953 to 1960, GDP grew at an average annual rate in excess of 7 percent.[3]

Economic growth accelerated in the early 1960s. Between 1960 and

3. Council for Economic Planning and Development, *Taiwan Statistical Data Book, 1980* (Taipei: CEPD, 1980), pp. 21–26.

1979, Taiwan's GDP increased at an average annual rate of nearly 10 percent, about 1.5 percentage points faster than that of Japan. In 1979 GDP per capita was approximately $1,500 (in 1979 New Taiwan dollars converted at a rate of 36 to the dollar). Data for Taiwan's population, GDP, and final energy consumption are given in table 3-8 for the period 1960–79.

Taiwan's population grew almost 3 percent a year during the 1960s and about 2 percent a year in the 1970s. For the entire period 1960–79, the average annual rate of population growth was 2.6 percent. As a consequence of this rapid growth of population, GDP per capita increased at an average annual rate of 7 percent.

The structure of Taiwan's economy changed greatly in the 1960s and 1970s, as shown by a comparison of the percentage shares of principal sectors contributing to real GDP in 1961 and 1979:[4]

	1961	1979
Agriculture	29.0	9.1
Manufacturing	19.9	43.2
Commerce	26.0	26.8
Transportation	2.9	6.0
Government services	15.8	8.4
Other	6.4	6.5

The most striking changes—again, typical of rapidly developing economies—were the sharp fall in the share of agriculture and the large increase in the share of manufacturing.

Rapid economic growth was associated with an even faster increase in final energy consumption,[5] which grew at an average annual rate of 10.6 percent from 1960 to 1979. The ratio of energy use to GDP increased at an average annual rate of 0.8 percent. Almost all of this increase took place after 1973, an unexpected occurrence in view of the huge increase in crude oil prices that accompanied the 1973–74 oil crisis.

That crisis does not appear to have been the kind of turning point for economic growth and energy consumption in Taiwan that it was in Japan. The growth rate of Taiwan's GDP did fall sharply in 1974, and it was far below normal in 1975. From 1976 to 1978, however, GDP resumed the

4. Figures supplied by the Energy Committee of the Ministry of Economic Affairs (Taipei). "Agriculture" includes forestry, livestock, fishing, and hunting. "Commerce" includes wholesale and retail trade, banking, insurance, real estate, restaurants, and other services. "Transportation" includes storage and communications.

5. Final energy consumption does not include transformation losses. The data used here also do not include "noncommercial" energy (mostly wood and charcoal).

Table 3-8. *Population, GDP, and Final Energy Consumption in Taiwan, 1960–79*

Year	Population (thousands)	Population growth rate (percent)	GDP (billions of 1976 New Taiwan dollars)	GDP growth rate (percent)	Final energy consumption (trillions of kilocalories)[a]	Energy consumption growth rate (percent)	GDP/population (thousands of 1976 New Taiwan dollars per capita)	Energy/population (millions of kilocalories per capita)	Energy/GDP (kilocalories per 1976 New Taiwan dollar)
1960	10,792	3.46	161.0	6.45	27.1	7.54	14.92	2.51	168.32
1961	11,149	3.31	172.1	6.89	28.7	5.90	15.44	2.57	166.76
1962	11,512	3.26	185.6	7.88	31.5	9.76	16.12	2.74	169.72
1963	11,884	3.23	203.0	9.38	32.5	3.17	17.08	2.73	160.10
1964	12,257	3.14	227.8	12.22	37.0	13.85	18.59	3.02	162.42
1965	12,628	3.03	253.2	11.15	40.2	8.65	20.05	3.18	158.77
1966	12,993	2.89	275.8	8.93	45.7	13.68	21.23	3.52	165.70
1967	13,297	2.34	305.3	10.70	50.5	10.50	22.96	3.80	165.41
1968	13,650	2.65	333.1	9.11	59.0	16.83	24.40	4.32	177.12
1969	14,335[b]	5.02[b]	362.7	8.89	63.3	7.29	25.30[b]	4.42	174.52

1970	14,676	2.38	403.8	11.33	69.7	10.11	27.51	4.75	172.61
1971	14,995	2.17	455.4	12.77	76.4	9.61	30.37	5.10	167.76
1972	15,289	1.96	515.7	13.24	86.8	13.61	33.73	5.68	168.31
1973	15,565	1.81	582.1	12.87	98.2	13.13	37.40	6.31	168.70
1974	15,852	1.84	588.7	1.13	97.5	−0.71	37.14	6.15	165.62
1975	16,150	1.88	616.9	4.79	108.5	11.28	38.20	6.72	175.88
1976	16,508	2.22	701.1	13.66	130.0	19.82	42.47	7.87	185.42
1977	16,813	1.85	769.7	9.78	142.3	9.46	45.78	8.46	184.88
1978	17,136	1.92	872.9	13.40	168.0	18.06	50.94	9.70	192.46
1979	17,479	2.00	940.6	7.76	183.8	9.40	53.81	10.52	195.41

Average annual growth rate

1960–73	...	2.86	...	10.39	...	10.41	...	...	...
1973–79	...	1.95	...	8.33	...	11.01	...	...	...
1960–79	...	2.57	...	9.74	...	10.60	...	...	...

Sources: Population: Directorate-general of Budget, Accounting, and Statistics, Executive Yuan, *Statistical Yearbook of the Republic of China, 1980* (Taipei: Veterans Printing Works, 1980), p. 2. GDP: Council for Economic Planning and Development, *Taiwan Statistical Data Book, 1982* (Taipei: CEPD, 1982), p. 21. Energy: Ministry of Economic Affairs, Energy Committee, *Taiwan Energy Statistics, 1979* (Taipei: MEA, 1979), pp. 26–31, 80–83, 88–89. Numbers are rounded.

a. Final energy consumption was calculated from energy balance sheets by using a constant electricity conversion ratio of 860 kilocalories per kilowatt hour. Noncommercial energy (mostly wood and charcoal) is not included. A trillion kilocalories approximately equal a hundred thousand t.o.e.

b. Population figures for 1969–79 include servicemen living on base and prison inmates not included in 1960–68 figures.

Table 3-9. *Energy Balance in Taiwan, 1960*
Ten billions of kilocalories[a]

Item	Coal[b]	Oil[c]	Natural gas	Hydro-electric[d]	Total elec-tricity[e]	Total
Primary energy supply						
Domestic production	2,575.2	2.1	22.9	508.1	0.0	3,108.3
Net imports	− 170.0	847.1	0.0	0.0	0.0	677.1
Stock changes	19.4	15.4	0.0	0.0	0.0	34.8
Total	2,424.6	864.6	22.9	508.1	0.0	3,820.2
Transformation sector	− 706.0	− 150.6	− 3.7	− 508.1	256.0	− 1,112.4
Final energy consumption, by sector						
Industrial	1,148.4	206.1	6.2	0.0	191.4	1,552.1
Residential and commercial	358.3	348.4	6.6	0.0	63.9	777.2
Transportation	211.9	107.4	1.4	0.0	0.7	321.4
Nonenergy uses of fuels	0.0	52.1	5.0	0.0	0.0	57.1
Total	1,718.6	714.0	19.2	0.0	256.0	2,707.8

Source: Based on more detailed data in Ministry of Economic Affairs, *Taiwan Energy Statistics, 1980*, pp. 52–53, 106–07. Numbers are rounded.

a. Ten billion kilocalories approximately equal a thousand t.o.e.

b. Includes coal, coke, gasworks gas, and patent fuel (briquets).

c. Includes crude oil, liquefied petroleum gas, aviation gasoline, motor gasoline, jet fuel, kerosene, diesel oil, fuel oil, and other petroleum products.

d. Electricity generated by hydropower (and when applicable, by nuclear power) has been assigned a generating efficiency of 35.1 percent.

e. No electricity was generated by nuclear plants until 1979.

high rate of growth that marked the years before the crisis. Final energy consumption followed a roughly similar path, although its drop in 1974 was more precipitous, and it recovered to the levels before the crisis more quickly than did GDP. Both GDP and energy consumption fell again in the 1979 oil crisis, but not to the extent that they did in 1974.

Simplified energy balances for Taiwan are presented for 1960, 1973, and 1979 (tables 3-9 through 3-14). To show structural changes more clearly, these balances have also been converted to percentages.

In 1960 Taiwan's total primary energy supply (not including wood and charcoal) was the equivalent of about 75,000 barrels of oil a day. By 1979 total primary energy supply had increased to the equivalent of over 500,000 barrels of oil a day. There was no possibility of obtaining this amount of energy from Taiwan's limited domestic resources. In 1960 about 80 percent of Taiwan's energy was domestic, but in 1979 approximately the same percentage was imported (if nuclear power is regarded as imported). The most fundamental change, most of which took place from 1960 to 1973, was the substitution of imported oil for domestic coal.

Table 3-10. *Structure of Energy Supply and Disposition in Taiwan, 1960*
Percent[a]

Item	Coal[b]	Oil[c]	Natural gas	Hydro-electric[d]	Total elec-tricity[e]	Total
Primary energy supply						
Domestic production	67.4	0.1	0.6	13.3	0.0	81.4
Net imports	−4.5	22.2	0.0	0.0	0.0	17.7
Stock changes	0.5	0.4	0.0	0.0	0.0	0.9
Total	63.4	22.7	0.6	13.3	0.0	100.0
Disposition of energy supply, by sector						
Transformation	18.5	3.9	0.1	13.3	−6.7	29.1
Industrial	30.1	5.4	0.2	0.0	5.0	40.7
Residential and commercial	9.4	9.1	0.2	0.0	1.7	20.4
Transportation	5.5	2.8	*	0.0	*	8.3
Nonenergy uses of fuels	0.0	1.4	0.1	0.0	0.0	1.5
Total	63.5	22.6	0.6	13.3	0.0	100.0

Source: Table 3-9.
* Less than 0.05 percent.
a. Numbers are rounded.
b–e. See notes to table 3-9.

Table 3-11. *Energy Balance in Taiwan, 1973*
Ten billions of kilocalories[a]

Item	Coal[b]	Oil[c]	Natural gas	Hydro-electric[d]	Total elec-tricity[e]	Total
Primary energy supply						
Domestic production	2,062.8	151.3	1,354.9	833.0	0.0	4,402.0
Net imports	92.7	9,456.7	0.0	0.0	0.0	9,549.4
Stock changes	77.0	173.5	0.0	0.0	0.0	250.5
Total	2,232.5	9,781.5	1,354.9	833.0	0.0	14,201.9
Transformation sector	−212.5	−4,521.2	−309.6	−833.0	1,496.9	−4,379.4
Final energy consumption, by sector						
Industrial	1,770.9	2,443.7	662.9	0.0	1,009.0	5,886.5
Residential and commercial	156.9	1,107.1	73.2	0.0	483.1	1,820.3
Transportation	92.2	1,091.4	0.0	0.0	4.8	1,118.4
Nonenergy uses of fuels	0.0	618.1	309.2	0.0	0.0	927.3
Total	2,020.0	5,260.3	1,045.3	0.0	1,496.9	9,822.5

Source: Based on more detailed data in Ministry of Economic Affairs, *Taiwan Energy Statistics, 1980*, pp. 78–79, 132–33.
a–e. See notes to table 3-9.

Table 3-12. *Structure of Energy Supply and Disposition in Taiwan, 1973*

Percent[a]

Item	Coal[b]	Oil[c]	Natural gas	Hydro-electric[d]	Total electricity[e]	Total
Primary energy supply						
Domestic production	14.5	1.1	9.5	5.9	0.0	31.0
Net imports	0.7	66.6	0.0	0.0	0.0	67.3
Stock changes	0.5	1.2	0.0	0.0	0.0	1.7
Total	15.7	68.9	9.5	5.9	0.0	100.0
Disposition of energy supply, by sector						
Transformation	1.5	31.8	2.2	5.9	− 10.5	30.9
Industrial	12.5	17.2	4.7	0.0	7.1	41.5
Residential and commercial	1.1	7.8	0.5	0.0	3.4	12.8
Transportation	0.6	7.7	0.0	0.0	*	8.3
Nonenergy uses of fuels	0.0	4.4	2.2	0.0	0.0	6.6
Total	15.7	68.9	9.6	5.9	0.0	100.0

Source: Table 3-11.
* Less than 0.05 percent.
a. Numbers are rounded.
b–e. See notes to table 3-9.

Table 3-13. *Energy Balance in Taiwan, 1979*

Ten billions of kilocalories[a]

Item	Coal[b]	Oil[c]	Natural gas	Nuclear[d]	Hydro-electric[d]	Total electricity[f]	Total
Primary energy supply							
Domestic production	1,686.3	207.5	1,704.0	1,551.2	1,119.2	0.0	6,268.2
Net imports	1,414.7	18,394.6	0.0	0.0	0.0	0.0	19,809.3
Stock changes	− 38.7	789.1	0.0	0.0	0.0	0.0	750.4
Total	3,062.3	19,391.2	1,704.0	1,551.2	1,119.2	0.0	26,827.9
Transformation sector	− 1,231.0	− 7,312.3	− 199.9	− 1,551.2	1,119.2	2,967.4	− 8,446.2
Final energy consumption, by sector							
Industrial	1,586.4	5,799.6	558.3	0.0	0.0	1,984.2	9,928.5
Residential and commercial	240.3	1,557.8	281.3	0.0	0.0	959.4	3,038.8
Transportation	4.6	2,553.2	0.0	0.0	0.0	23.8	2,581.6
Nonenergy uses of oil	0.0	2,168.3	664.5	0.0	0.0	0.0	2,832.8
Total	1,831.3	12,078.9	1,504.1	0.0	0.0	2,967.4	18,381.7

Source: Based on more detailed data in Ministry of Economic Affairs, *Taiwan Energy Statistics, 1980*, pp. 90–91, 144–45.
a–e. See notes to table 3-9.

Table 3-14. *Structure of Energy Supply and Disposition in Taiwan, 1979*
Percent[a]

Item	Coal[b]	Oil[c]	Natural gas	Nuclear[d]	Hydro-electric[d]	Total electricity[e]	Total
Primary energy supply							
Domestic production	6.3	0.8	6.4	5.8	4.2	0.0	23.5
Net imports	5.3	68.6	0.0	0.0	0.0	0.0	73.9
Stock changes	−0.1	2.9	0.0	0.0	0.0	0.0	2.8
Total	11.5	72.3	6.4	5.8	4.2	0.0	100.0
Disposition of energy supply, by sector							
Transformation	4.6	27.3	0.7	5.8	4.2	−11.1	31.5
Industrial	5.9	21.6	2.1	0.0	0.0	7.4	37.0
Residential and commercial	0.9	5.8	1.0	0.0	0.0	3.6	11.3
Transportation	*	9.5	0.0	0.0	0.0	0.1	9.6
Nonenergy uses of fuels	0.0	8.1	2.5	0.0	0.0	0.0	10.6
Total	11.4	72.3	6.3	5.8	4.2	0.0	100.0

Source: Table 3-13.
* Less than 0.05 percent.
a. Numbers are rounded.
b–e. See notes to table 3-9.

In 1960 domestic coal supplied two-thirds of Taiwan's energy; in 1973 imported oil supplied two-thirds. Over 90 percent of Taiwan's energy imports (not counting nuclear fuel) are in the form of oil, most of it from the Persian Gulf.

The greatest changes in the disposition of Taiwan's primary energy supply during the period 1960–79 were the sharp drop in the share of the residential and commercial sector and the large increase in the share of the nonenergy-uses sector. The industrial sector experienced a moderate decline in its share, and the shares of the transformation and transportation sectors rose slightly.

The overall intensity of energy use (measured by the ratio of total energy use, including transformation losses, to GDP) increased very little before 1973. In 1979, however, it was 15.8 percent greater than it had been in 1973. If the ratio of energy use to GDP had remained at the 1973 level, Taiwan's 1979 GDP could have been produced with 14.4 percent less energy. The transformation and nonenergy-uses sectors each were responsible for about 35 percent of the overall increase in energy intensity.

The thermal efficiency of energy use outside the transformation sector increased sharply from 1960 to 1973, but very little from 1973 to 1979.

The rise in thermal efficiency was partly the result of changes in the relative sizes of the energy-consuming sectors. The primary cause, however, was the shift from coal to oil, gas, and electricity.

The rise in thermal efficiency reduced the rate of increase in the overall energy intensity of Taiwan's economy. Two other general developments worked in the opposite direction. First, transformation losses increased from 29.1 percent of primary energy supply in 1960 to 31.5 percent in 1979. Second, the export surplus of energy embodied in Taiwan's nonfuel international trade rose from 9.8 percent of total primary energy supply in 1973 to 15.7 percent in 1979. The rise in transformation losses was attributable to the increase in the proportion of total primary energy consumed in the form of electricity and refined petroleum products. The increase in the net export of embodied energy was the result of a widening surplus of nonfuel exports over nonfuel imports.

Real energy prices increased about 65 percent from 1973 to 1978 and then declined slightly in 1979 to 64 percent above the 1973 level. Over half of this increase took place in 1974. The rise in real energy prices would have been greater if the government had not pursued a policy, in response to the 1973–74 and 1979–80 oil crises, of deliberately increasing the general price level to depress consumer purchasing power.

The increase in energy prices undoubtedly moderated the growth rate of energy consumption. Energy consumption per capita, however, grew 66.7 percent from 1973 to 1979, whereas GDP per capita grew 43.9 percent. The effect of higher prices was therefore not strong enough to overcome other developments that increased the intensity of energy use.

Until fairly recently, the government appears not to have tried to influence the energy intensity of the economy. Its main concern has been to ensure the increasing amounts of energy needed to sustain rapid economic growth. Especially during the 1970s, government development policies contributed significantly to the intensity of energy use. In January 1979, however, the cabinet approved a comprehensive statement of energy policy that dealt with energy conservation as well as with problems of energy supply. In 1980 the legislature passed an energy management law that was promulgated by the president on August 8 of that year. Important provisions of this law give the government specific authority to set energy conservation standards and to establish penalties for noncompliance.

Comparison of Developments in South Korea and Taiwan

During the 1960s and 1970s, both South Korea and Taiwan experienced rapid economic growth that greatly increased energy consumption. Because known domestic energy resources in both countries are limited, Korea's dependence on imported energy rose from 11.7 percent in 1965 to 79.5 percent in 1979, and Taiwan's energy import dependence increased from 28.7 percent to 73.9 percent over the same period.

Despite similarities in their recent economic and energy developments, the two economies have major structural differences. In the early 1960s Korea was essentially an agricultural society, and by the late 1970s agriculture still contributed a fifth of the country's GDP. In Taiwan, agriculture accounted for less than 30 percent of GDP in 1961 and less than 10 percent in 1979. Throughout the 1960s and 1970s, manufacturing has been less important in the economy of Korea than it has been in the economy of Taiwan, although the gap between the two has been narrowing. Because the two economies are structurally different, it is not surprising that the structures of their energy supply and disposition have also not been the same.

In 1965 Korea obtained over 40 percent of its energy from noncommercial fuels. Information on the use of such fuels in Taiwan is lacking, but it appears unlikely that they could have accounted for as much as 10 percent of total energy use in the mid-1960s. Coal was the most important commercial fuel in both countries in 1965, although oil already provided over one-fourth of Taiwan's energy. During the late 1960s and 1970s, oil replaced coal as the leading fuel in both Korea and Taiwan. This process did not go as far in Korea, however, as it did in Taiwan, largely because of the continued importance of coal in heating Korean homes. Throughout the period, Taiwan enjoyed the advantage of significant supplies of domestic natural gas and larger developed hydropower resources.

The most striking difference between Korea and Taiwan in the disposition of total primary energy supply is in the relative importance of residential, commercial, and miscellaneous energy consumption in total energy use (table 3-15). In 1965 these uses ("other" in table 3-15) accounted for 69.5 percent of Korea's energy use, and in 1979 their share of total use was still 43.1 percent. The comparable figures for Taiwan in 1965 and 1979 were 25.0 percent and 19.5 percent. This difference can

Table 3-15. *Structure of Disposition of Energy Supply*
in South Korea and Taiwan, by Sector, 1965, 1973, and 1979
Percent[a]

	1965		1973		1979	
Sector[b]	Korea	Taiwan	Korea	Taiwan	Korea	Taiwan
Industrial[c]	24.6	65.4	29.7	66.2	44.3	63.9
Transportation	5.9	9.6	15.5	13.4	12.6	16.6
Other[d]	69.5	25.0	54.8	20.5	43.1	19.5
Total	100.0	100.0	100.0	100.0	100.0	100.0

Sources: Tables 3-3, 3-5, 3-7, 3-12, and 3-14; 1965 figures for Taiwan were derived from Ministry of Economic Affairs, *Taiwan Energy Statistics, 1980*, pp. 116–17.

a. Numbers are rounded.

b. Transformation losses in Taiwan have been distributed on a proportional basis to the three consuming sectors: "industrial," "transportation," and "other." The statistical difference between primary energy supply and final energy consumption in Korea has been similarly distributed among the consuming sectors. Data for Korea do not show transformation losses for coal products and petroleum products separately, and it is assumed that these losses have also been distributed among the consuming sectors. Data for Taiwan do not include noncommercial energy, but data for Korea do.

c. "Industrial" includes nonenergy coal and coal products. Petroleum products and natural gas used as nonenergy materials are not included because they are not recorded in Korean energy statistics.

d. "Other" includes residential, commercial, public, and miscellaneous uses.

be explained by Korea's lower level of industrialization and colder climate, as well as by the fact that Taiwan does not collect information on noncommercial fuels, which are largely consumed in this sector.

As might be expected, the industrial sector was more important in Taiwan than in Korea. It already accounted for 65.4 percent of total energy use in 1965, but its share declined to 63.9 percent in 1979 after having increased to 66.2 percent in 1973. The share of Korea's industrial sector increased from 24.6 percent to 44.3 percent over the same period, with most of the increase taking place after 1973.

The overall energy intensities of the two economies, measured by the ratio of total energy use to GDP, followed different paths. From 1965 to 1973, energy intensity in Korea fell by about 2 percent, whereas it rose nearly 10 percent in Taiwan. After 1973 energy intensity in Taiwan increased, but it continued to fall in Korea. In 1979 energy intensity in Taiwan was 28.5 percent greater than it had been in 1965. Over the same period, energy intensity in Korea decreased 8.7 percent.

The fall in energy intensity in Korea compared with the rise in Taiwan is the result of several factors. The thermal efficiency of energy use increased more rapidly in Korea than it did in Taiwan, an increase that allowed Korea to consume more useful energy without a commensurate rise in the consumption of gross energy.[6] Transformation losses were

6. Thermal efficiency in Korea, however, was lower than in Taiwan throughout the period 1965–79.

lower in Korea than in Taiwan because smaller percentages of primary energy supply were consumed in the form of electricity and refined petroleum products. The energy content of nonfuel international trade also contributed to Korea's relative advantage in energy intensity. Korea was a net importer of embodied energy, whereas Taiwan was a net exporter. Finally, as the consequence of opposite movements of the exchange rate between national currencies and the U.S. dollar, domestic energy prices rose in Korea but fell in Taiwan during the period 1965–73.[7]

In sectoral terms, the small decrease in the overall intensity of the Korean economy from 1965 to 1973 was caused by a fall in the ratio of energy use to GDP in the "other" sector, and the drop in intensity from 1973 to 1979 was the result of a decrease in this ratio within the transportation and "other" sectors. The increase in energy intensity in Taiwan in both periods is attributable principally to a rise in the ratio of energy use to GDP in the industrial sector.

The governments of Taiwan and South Korea have played generally similar roles in influencing energy developments. Both governments have promoted rapid economic growth that has increased energy consumption, and both have made developmental decisions that have contributed to rises in energy intensity. The first oil crisis caused both governments to try to diversify energy sources in order to reduce dependence on imported oil. Both governments moved more slowly in the field of energy conservation, but they now have well-developed conservation programs, especially in the industrial sector.

7. Energy prices rose about 10 percent faster in Korea than in Taiwan from 1973 to 1980.

CHAPTER FOUR

Adjustments to the Oil Crises of 1973–74 and 1979–80

ON OCTOBER 6, 1973, Arab forces launched a surprise attack on Israeli forces along the Suez Canal and on the Golan Heights. Although a truce was reestablished in only a few weeks, the repercussions of the October 1973 war on international energy markets were profound and lasting.[1]

When the war broke out, the price of 34° Arabian light crude oil (the benchmark or marker crude), free on board (f.o.b.) the Persian Gulf, was approximately $2.70 a barrel. On October 16, taking advantage of fears concerning the security of oil supplies, the Persian Gulf oil-exporting countries raised the price of oil to $3.65 a barrel. Other exporters quickly adopted the new price.

Five days later, the Arab oil-exporting countries (except Iraq) cut oil production by 5 or 10 percent and threatened deeper cuts unless their objectives with respect to Israel were achieved. At the same time, all of the Arab oil-exporting countries (including Iraq) imposed an embargo on oil shipments to the United States, the Netherlands, and several other countries that they regarded as too pro-Israeli. Several other countries, including the United Kingdom and France, were designated "friendly" to the Arab cause and assured normal oil supplies. Japan, South Korea, and Taiwan were among the countries that were neither embargoed nor

1. The summary of developments in late 1973 and early 1974 presented here is drawn from Joseph A. Yager and Eleanor B. Steinberg, "Trends in the International Oil Market," in Edward R. Fried and Charles L. Schultze, eds., *Higher Oil Prices and the World Economy: The Adjustment Problem* (Brookings Institution, 1975), especially pp. 230–39.

labeled friendly. Such countries were by implication vulnerable to unspecified reductions in their normal supplies of imported Arab oil.

On November 4, participating Arab countries announced a further cut in production to 75 percent of the level produced in September 1973. In late December, however, production was raised to 85 percent of the September level. Total Arab production actually fell to about 80 percent of the September level in both November and December.[2] In March, the Arab oil-exporting countries removed both the embargo and production restrictions.

Although the Arab production restrictions did not last long and never did cut very deep,[3] they facilitated additional and much larger price increases. Effective January 1, 1974, the Persian Gulf oil-exporting countries more than doubled the posted price, thereby raising the market price to $9.66 a barrel. Other exporters soon announced comparable increases, and further smaller increases pushed the price to $10.47 a barrel at the end of 1974. A second oil crisis occurred in 1979–80. The revolution in Iran caused a tightening of the international oil market and a large increase in crude oil prices. The price of Saudi Arabian light, the marker crude, rose from $12.70 a barrel in December 1978 to $30.00 a barrel in December 1980.[4] Roughly 30 percent of this increase took place in 1979 and the remainder in 1980.

Adjustment of Japan

Early Japanese reactions to the October 1973 war and subsequent events focused on the threat to the physical volume of Japan's oil imports from the Middle East. Only later did adjusting to higher oil prices become the primary concern.[5]

2. Federal Energy Administration, Petroleum Industry Monitoring System, *U.S.-OPEC Petroleum Report, Year 1973* (FEA, July 1, 1974).

3. Ibid. At their peak in November and December 1973, these restrictions cut production about 3.4 million barrels a day below the September level. This was about 7 percent of crude oil production in non-Communist countries.

4. "Back to OPEC's Long-Term Strategy," *Petroleum Economist*, vol. 48 (December 1981), p. 515.

5. For accounts of Japanese reactions to the oil embargo, see Tsunehiko Watanabe, "Japan," in Fried and Schultze, eds., *Higher Oil Prices*, pp. 143–67; and Joseph A. Yager and Eleanor B. Steinberg, *Energy and U.S. Foreign Policy* (Ballinger, 1974), pp. 149–51.

In 1972, 43 percent of Japan's crude oil imports had come from Arab countries.[6] Although Japan was not embargoed, neither had it been designated a friendly country. The Japanese feared that the largely American international oil companies that handled most of the country's oil imports would discriminate against Japan and that Japan would have to bear a disproportionate share of the Arab reductions of oil production. In the actual event, the companies tried to spread the reductions evenly by shifting non-Arab oil to customers who were denied all or part of their normal Arab supplies.

The impact of the embargo on Japan's oil imports cannot be measured precisely because there is no way of knowing how much oil Japan would have imported under different circumstances. Crude oil imports in 1974 were 4 percent below those in 1973. But oil imports in the first quarter of 1974, when (allowing for transit time) the direct effect of the embargo should have been at its peak, were slightly larger than they were in the same quarter of 1973.[7] The principal impact of the embargo on the volume of Japan's oil imports was probably on demand rather than on supply. By checking the growth of the gross national product (GNP), the higher oil prices that the embargo made possible reduced oil consumption indirectly as well as directly.

Changes in Japan's oil stocks provide further evidence that the embargo did not seriously curtail the availability of oil below the amounts demanded. End-of-month oil stocks in millions of barrels were: September 1973, 306; December 1973, 281; March 1974, 260; June 1974, 331; September 1974, 365; December 1974; 330.[8] The fall in stocks in the last quarter of 1973 and first quarter of 1974 may have been caused partly by the embargo, but it was also a normal seasonal phenomenon.

Whatever its effect on the availability of oil, the embargo did not last long. Japan worked hard to improve its relations with the Arab countries and adopted a more pro-Arab position with respect to the Arab-Israeli dispute. On December 25, 1973, the Arabs rewarded Japan by declaring it a friendly nation. This diplomatic success did not, of course, remove the continuing problem of adjusting to greatly increased oil prices.

Higher prices of imported oil affected the economies of Japan and

6. Organization for Economic Cooperation and Development, International Energy Agency (IEA), *Oil Statistics: Supply and Disposal* (Paris: OECD, 1973), pp. 26–27.
7. U.S. Central Intelligence Agency, Office of Economic Research, *International Oil Developments: Statistical Survey*, ER-IOD-SS-76-003 (CIA, April 22, 1976), p. 9.
8. Ibid., p. 18.

other oil-importing countries in several ways. Since the demand for oil is inelastic, the oil import bill was increased. The general price level was pushed up, both because of the increased price of petroleum products and because of the increased cost of commodities that contain oil or whose production requires the use of oil as energy. In addition, the overall level of economic activity was depressed because consumers who were required to pay more for petroleum products and products dependent on oil had less to spend on other things.[9] All of these effects of higher oil prices were to some extent undesirable. How undesirable depended in part on the state of the economy of an oil-importing country at the time of the price increase.

In the case of Japan, the oil crisis occurred at a time of accelerating inflation.[10] The wholesale price index (WPI) in January 1973 was 7.6 percent above its level in January 1972. Prices rose steadily through the first three quarters of the year. By September the WPI was 18.7 percent higher than it had been a year previously. In July and August, the Bank of Japan acted to cool the overheated economy by raising the official discount rate, increasing required reserve ratios, and applying informal administrative controls on commercial bank lending. Fiscal policy, however, remained expansionary. A high rate of economic growth was expected to continue through 1974, and the problem of controlling inflation was expected to become more serious.

Under these circumstances the Japanese government decided to give priority to countering the inflationary effects of higher oil prices at the cost of accepting, and even deepening, the depressing effects of those prices on the level of economic activity. In late December 1973 direct government controls were imposed on the prices of major commodities and on the consumption of oil and electricity. The Bank of Japan again raised the rediscount rate and used informal administrative controls on bank lending to reduce the money supply. Plans for government spending in the fiscal year beginning April 1, 1974, were cut back significantly.

The war and the embargo had initially stimulated panic buying that

9. Income was also transferred from oil-importing to oil-exporting countries, but there was nothing that the importing countries could do to prevent this consequence of higher oil prices.

10. The account of developments in Japan immediately before and after the oil embargo is based largely on Watanabe's chapter in Fried and Schultze, eds., *Higher Oil Prices,* and on OECD, *OECD Economic Surveys: Japan* (Paris: OECD, July 1974, July 1975).

made inflation worse. By January 1974, however, consumer attitudes were more calm, and the government was able to plan the phased removal of emergency price controls by September. At the end of March, an average increase of 62 percent in the price of petroleum products was authorized. Other increases in commodity prices were permitted in following months.

Before the embargo and the jump in oil prices, Japan's real GNP was expected to increase more than 5 percent in 1974. It actually declined by about 1 percent.[11] The total shortfall of about 6 percent was the combined consequence of higher oil prices and the measures adopted by the Japanese government to counter the inflationary effects of those higher prices.

Negative economic growth in 1974 and slower growth in subsequent years inevitably were reflected in energy consumption. The effect of the crisis on energy consumption was increased, however, because—with a short delay—higher crude oil prices caused increases in the prices of petroleum products. The wholesale index of prices of petroleum, coal, and their products rose 128 percent from 1973 to 1974.[12] From October 1973 to January 1975 the retail prices of regular gasoline, diesel fuel, and heating oil increased 54, 55, and 121 percent, respectively. If taxes are excluded, the prices of regular gasoline and diesel fuel increased 83 and 90 percent, respectively (heating oil was not taxed).[13]

The effect of higher oil prices on Japan's balance of trade was remarkably transitory (top part of table 4-1). After experiencing a favorable trade balance in 1973, Japan was thrown into deficit in the first quarter of 1974 by the sharp increase in oil prices. In the remaining quarters of 1974, Japan's exports outpaced its imports and produced increasing trade surpluses.

The stagnation of imports during most of 1974 reflected the generally depressed state of the economy and can therefore be credited in part to the government's deflationary policies. Higher oil prices worked both to hold down the general demand for imports and to curtail the demand for imported oil in particular. The weakness of domestic demand may have contributed to the increase in Japan's exports, but a more obvious

11. *OECD Economic Surveys: Japan* (Paris: OECD, July 1975), p. 63. A decline of 1.8 percent was assumed in Fried and Schultze, eds., *Higher Oil Prices*, p. 21.

12. Prime Minister's Office, Statistics Bureau, *Japan Statistical Yearbook, 1980* (Tokyo: Statistics Bureau, 1980), p. 374.

13. U.S. Central Intelligence Agency, *International Oil Developments*, April 22, 1976, p. 20.

Table 4-1. *Effect of Higher Oil Prices on Japan's Balance of Trade, 1973–74 and 1978–80*
Billions of dollars, seasonally adjusted

Year and quarter	Exports	Imports	Balance
First oil crisis			
1973	54.5	53.0	1.5
1974	54.1	53.0	1.1
I	11.2	12.3	− 1.1
II	13.7	13.6	0.1
III	14.2	13.5	0.7
IV	15.0	13.6	1.4
Second oil crisis			
1978	95.6	71.0	24.6
1979	101.2	99.4	1.8
1980	126.7	124.6	2.1
I	28.2	30.0	− 1.8
II	31.4	32.1	− 0.7
III	32.3	30.7	1.6
IV	34.8	31.9	2.9

Sources: For 1973–74, Organization for Economic Cooperation and Development, *OECD Economic Surveys: Japan* (Paris: OECD, July 1975), p. 31. For 1978–80, *OECD Economic Surveys: Japan* (Paris: OECD, July 1981), p. 20. Numbers are rounded.

explanation is the relative strength of demand for the particular commodities exported by Japan and in the geographic areas to which Japan exported.

During the second oil crisis, energy prices to final users in Japan rose 13.3 percent in 1979 and 59.2 percent in 1980.[14] The pass-through ratio calculated by the Organization for Economic Cooperation and Development (OECD) for 1978–80 was 0.57, substantially higher than the ratio of 0.35 calculated for 1973–75.[15]

In gross terms, Japan's successful adjustment to the 1979–80 increases in energy prices is indicated by the following index numbers (1978 = 100):[16]

14. "Prospects for Recovery," *OECD Observer*, no. 108 (January 1981), p. 17.

15. Ibid., p. 16. A pass-through ratio is the change in energy prices to final users relative to the rise in the import price of oil.

16. The index of wholesale fuels and energy prices is from Prime Minister's Office, Statistics Bureau, *Japan Statistical Yearbook, 1982,* p. 399. The index of final energy consumption per capita is based on energy balances by the Institute of Energy Economics (Japan) and population data in *Japan Statistical Yearbook, 1982,* p. 15. The GNP figures used in the index of GNP per capita are from *OECD Economic Surveys, 1981–82: Japan* (Paris: OECD, July 1982), p. 67.

	Wholesale fuels and energy prices	Final energy consumption per capita	GNP per capita
1977	106.1	99.3	95.9
1979	109.2	103.9	104.4
1980	179.9	101.6	107.8
1981	201.3	99.8	110.3

The relatively modest rise in energy prices in 1979 was not sufficient to prevent a small increase in energy consumption per capita. The much larger increase in energy prices in 1980 did, however, cause energy consumption per capita to fall 2.2 percent. The most remarkable feature of the figures above is the continued rise in GNP per capita despite the depressing effect of higher energy prices on domestic demand.

The overall energy intensity of the Japanese economy (measured by the ratio of total energy use to GNP) declined only slightly (0.4 percent) in 1979. In 1980 and 1981, however, it declined 5.4 percent and 4.9 percent, respectively.

As was the case in 1973–74, the effects of increased oil prices in 1979–80 on Japan's balance of trade were remarkably brief (bottom part of table 4-1). The favorable trade balance shrunk in 1979 because imports increased more rapidly than exports. Oil accounted for only about a third of the increase in imports. In 1980 imports and exports increased by about the same amount, with oil accounting for about two-thirds of the increase in imports.[17]

The increase in international oil prices caused a substantial temporary deterioration in Japan's terms of trade. Export prices increased 8.9 percent in 1979 and 9.3 percent in 1980, but import prices rose 28.6 percent in 1979 and 43.0 percent in 1980. Quarterly figures, however, show an improvement in the terms of trade in the course of 1980 and early 1981.[18]

The exchange rate of the yen against the U.S. dollar and other major currencies fluctuated widely during the 1979–80 oil crisis. From October 1978 to November 1979, the yen depreciated about 25 percent. It began to recover in April 1980, and by February 1981 it had appreciated about 28 percent. The changes in the value of the yen are attributable in large part to a sizable deficit on current account in 1979 and early 1980 followed by a shrinking of that deficit in the remainder of 1980 and early 1981.[19]

17. OECD Economic Surveys: Japan (Paris: OECD, July 1981), p. 23.
18. Ibid., p. 16.
19. Ibid., pp. 20, 26, 27.

The lagged effect of the depreciation of the yen was a major cause of the upsurge of exports in 1980.

The 1979–80 oil crisis caused wholesale prices to rise 7.3 percent in 1979 and 17.8 percent in 1980. Wholesale prices peaked, however, in the second quarter of 1980 and declined thereafter. In 1981 they increased only 1.7 percent. Consumer prices were much less affected by the oil crisis; they rose 3.6 percent in 1979, 8.0 percent in 1980, and 4.9 percent in 1981.[20]

Japan's successful adjustment to the second oil crisis was the combined result of fortunate circumstances and wise government policies.[21] The increase in oil prices was much more gradual in 1979–80 than in 1973–74. The economy therefore had more time to adjust, and the government had more time to apply anti-inflationary measures. Moreover, immediately before the crisis the economy was not overheated, as it had been in 1973. The rate of economic growth was moderate, prices were relatively stable, and excess productive capacity existed. During the first part of the 1979–80 crisis, the rapid depreciation of the yen facilitated an upsurge of exports that helped to compensate for the depressing effect of higher energy prices on domestic demand.

Structural changes in the manufacturing sector after the first oil crisis, which were encouraged by the government, reduced the vulnerability of the economy to the further increases in energy prices in the second oil crisis. Energy-intensive industries—such as pulp, paper and paper products, petroleum and coal products, nonmetallic mineral products, and basic metals—grew less rapidly than the manufacturing sector as a whole. Technology-intensive industries—such as chemicals, electrical machinery, transport equipment, and precision machinery—grew more rapidly.[22]

This structural change, along with a rationalization of production processes, reduced energy consumption per unit of output and increased labor productivity. From 1975 to 1980 labor productivity increased at an average annual rate of 7.3 percent while unit labor costs increased only 1.1 percent a year.[23] The strong increase in labor productivity and the

20. Ibid., p. 16; and Statistics Bureau, *Japan Statistical Yearbook, 1982*, pp. 398, 404.

21. The explanation of Japan's adjustment to the 1979–80 oil crisis presented here draws heavily on Walter R. Mahler, "Japan's Adjustment to the Increased Cost of Energy," *Finance & Development*, vol. 18 (December 1981), pp. 26–29.

22. *OECD Economic Surveys: Japan* (July 1981), pp. 45–46.

23. Ibid., p. 51.

moderate rise in labor costs made it easier for the government to contain the inflationary effects of the increases in oil prices.

The policy of the government appears to have been to reinforce through fiscal and monetary actions the economic forces that were working to moderate the impact of the oil price increases. The net effect of the government's fiscal policy was mildly deflationary in both 1979 and 1980.[24] Through a combination of informal guidance of bank lending and changes in the official rediscount rate and reserve requirements, the government also gradually reduced the annual rate of growth in the money supply from 12 percent in the first three quarters of 1979 to 7 percent in the first half of 1981.[25]

A comparison of the adjustments to the 1973–74 oil crisis of Japan, the United States, France, and West Germany is informative. The principal effect of the first oil crisis on these countries was not the temporary disruption of oil supplies but the fourfold increase in the price of crude oil. When the crisis unexpectedly erupted, all four countries were experiencing inflationary problems. These problems were aggravated by the oil crisis in all cases except West Germany (top part of table 4-2).

The initial reaction of all four governments was to maintain or to impose restrictive monetary and fiscal policies.[26] These policies added to the depressing effect of higher oil prices on aggregate demand and produced either a slowdown or an actual reversal of economic growth (bottom part of table 4-2). By 1975, restrictive policies had been, or were being, relaxed. By 1976, all four countries except Japan had come close to (or, in the case of West Germany, had exceeded) rates of economic growth before the oil crisis.

The effect of the 1973–74 oil crisis on total energy use in the four countries in 1974–77 can be expressed as index numbers (1973 = 100):[27]

	Japan	United States	France	West Germany
1974	100.8	98.2	96.2	97.7
1975	97.2	93.2	90.8	91.7
1976	102.0	99.1	96.2	99.2
1977	103.1	101.7	97.3	98.9

24. Ibid., p. 42.

25. Ibid., p. 30.

26. For a detailed account of the effects of the first oil crisis and governmental reactions to it, see Fried and Schultze, eds., *Higher Oil Prices.*

27. OECD, IEA, *Energy Balances of OECD Countries* (Paris: OECD, various issues).

Table 4-2. *Annual Increases in Consumer Prices and Annual Changes in Real Gross Domestic Product in Japan and Three Other Industrial Countries, 1972–77*
Percent

Year	Japan	United States	France	West Germany
Increase in consumer prices				
1972	4.5	3.3	6.2	5.5
1973	11.7	6.2	7.3	6.9
1974	24.5	11.0	13.7	7.0
1975	11.8	9.1	11.8	6.0
1976	9.3	5.8	9.6	4.5
1977	8.1	6.5	9.4	3.7
Change in real GDP				
1972	8.8	5.8	5.9	3.7
1973	8.8	5.4	5.4	4.9
1974	− 1.0	− 1.3	3.2	0.5
1975	2.3	− 1.0	0.2	− 1.8
1976	5.3	5.6	5.2	5.2
1977	5.3	5.1	2.8	3.0

Source: For consumer price increases, OECD, *OECD Economic Outlook,* no. 29 (Paris: OECD, July 1981), p. 140; for changes in real GDP, ibid., p. 132. Numbers are rounded.

The relatively shallow and brief dip in energy use in Japan compared with the other three countries corresponds to the similar behavior of Japan's gross domestic product, which in turn may reflect a government decision not to move too swiftly or drastically against inflation. The moderate impact of the first oil crisis on energy use in Japan may also be evidence of relatively low income and price elasticities of demand for energy.

In the years after the 1973–74 oil crisis all four countries tried to reduce their dependence on oil. Japan was the least successful in absolute terms. Japan's oil consumption was 16.8 percent greater in 1979 than it was in 1973, compared with a 9.2 percent increase in the United States and a 1.9 percent increase in France. West Germany's oil consumption in 1979 was slightly below (0.8 percent) its 1973 level.[28] Oil consumption was held down by reducing the share of oil in total primary energy supply. Reliance on gas and nuclear power (in the United States, on coal and nuclear power) reduced oil's share in total energy supply from 1973 to 1980 by 18.4 percent in France, 13.5 percent in West Germany, 14.5 percent in Japan, and 3.6 percent in the United States.

28. Ibid.

The consumption of oil in all four countries would have been higher during 1973–80 had there not been substantial decreases in the energy intensities of their economies.[29] Japan led with a 15.0 decrease. West Germany followed with a decrease of 13.1 percent. The United States and France had decreases of 11.6 and 12.0 percent, respectively. The decreases in energy intensity may have been brought about in part by the energy conservation programs that each of these countries adopted,[30] but most of the credit for the decreases probably must go to higher energy prices.

Energy prices increased much more in Japan than in the other three countries, and this explains Japan's lead in decreasing energy intensity. The sensitivity of energy intensity to increases in energy prices was lower in Japan than in the other countries. That is, a given percentage increase in energy prices was associated with a smaller percentage decrease in energy intensity.

The energy price policies of the four countries have tended to converge. West Germany has always relied mainly on market forces, although electricity prices are regulated (as they are in the other three countries), and coal prices are subsidized by the proceeds of a tax on electricity. The government of Japan can influence energy prices by administrative guidance but apparently intervenes in the market less in the early 1980s than it did in the immediate aftermath of the 1973–74 oil crisis. The United States has removed controls from domestic crude oil, and controls over natural gas prices are being phased out. France still regulates energy prices but passes price increases through to final consumers more quickly than it did immediately after the first oil crisis.

Pass-through ratios calculated by the OECD show that all four countries passed rises in the price of imported oil to final consumers more fully and promptly in the second than in the first oil crisis:[31]

	1973–75	1978–80
Japan	0.35	0.57
United States	0.23	0.50
West Germany	0.18	0.31
France	0.23	0.42

29. Because oil is the residual fuel in the current era, changes in energy intensity tend to be reflected directly in changes in oil consumption.

30. The annual reports of the IEA, *Energy Policies and Programmes of IEA Countries*, give information on the U.S., West German, and Japanese conservation programs.

31. "Prospects for Recovery," p. 16.

Not enough time has passed to permit a full comparison of the reactions of the four countries to the 1979–80 oil crisis.

The 1973–74 oil crisis and subsequent energy developments have affected the foreign policies of all four countries. Japan, West Germany, and the United States are members of the International Energy Agency, formed in 1974 in response to a U.S. initiative.[32] France has not joined the IEA but participates indirectly through the European Community.

All four countries have endeavored to improve their relations with the major oil-exporting countries. Particularly in the first year or two after the 1973–74 crisis, France, West Germany, and to a lesser extent Japan sought explicit or implicit bilateral agreements by which oil supplies would be assured in return for investments, technical assistance, and (in the case of France) arms sales. The United States opposed such arrangements as potentially damaging to unity among oil-importing countries. The United States, however, sought to make its supplies of imported oil more reliable by increasing its economic and security ties with the major oil-exporting countries.

Adjustments of Korea and Taiwan

Energy prices in South Korea and Taiwan rose substantially in both oil crises. The index numbers in table 4-3 provide a rough basis for comparing the changes in energy prices after 1973.

As in the earlier period, energy prices were affected by changes in the value of national currencies against the U.S. dollar, in which international oil prices are quoted. The Korean won depreciated 39.8 percent against the U.S. dollar from 1973 to 1980, whereas the New Taiwan dollar appreciated 5.7 percent against the U.S. dollar during the same period. These changes tended to push up energy prices in Korea and to moderate energy price increases in Taiwan. The indexes in table 4-3 suggest, however, that the effect of exchange rates on energy prices was submerged by other influences, including government intervention.

32. Functions of the IEA include planning for sharing of available oil supplies in emergencies, reviewing the energy policies and programs of member countries, sponsoring energy conservation projects, and disseminating energy statistics.

Table 4-3. *Index of Change in Energy Prices in Korea and Taiwan, 1974–80*
1973 = 100

Year	Korea	Taiwan
1974	146.3	132.5
1975	152.0	143.0
1976	147.3	139.6
1977	147.6	151.9
1978	145.3	149.6
1979	166.4	154.0
1980	216.3	197.1

Sources: Figures for Korea are the weighted average of the price indexes for petroleum and related products and for coal and electric power from Economic Planning Board, *Korea Statistical Yearbook, 1981* (Seoul: EPB, 1981), pp. 408–09. Figures for Taiwan are the index of water supply, electricity, and gas prices from Council for Economic Planning and Development, *Taiwan Statistical Data Book, 1981* (Taipei: CEPD, 1981), p. 168. Both series have been deflated by WPIs from the same sources. Numbers are rounded.

Korea

At the time of the first oil crisis, the Korean economy was showing definite signs of overheating. The GDP rose a phenomenal 14.7 percent in 1973. The money supply (M1, or currency in circulation plus deposit money) increased about 40 percent. Wholesale prices, however, increased only 6.9 percent.[33]

An increase in the rate of inflation would not have been surprising even without the huge rise in the price of imported oil. Wholesale prices actually increased 42 percent in 1974. The government does not appear to have taken any specific actions to deal with this upsurge in prices. Its fiscal and monetary policies certainly cannot be called restrictive. The surplus of current revenues over current expenditures was smaller in 1974 than it had been in 1973. Interest rates were by and large not pushed up, and the money supply grew another 30 percent.[34]

In retrospect the government appears to have regarded continued rapid economic growth as worth the price of increased inflation. It judged that the strong competitive position of Korean products in export markets would make it possible to pass a large part of the burden of higher oil prices to foreign consumers, thereby reducing the depressing effect of those prices on the Korean economy. The value of Korea's exports in

33. Economic Planning Board (EPB), *Korea Statistical Yearbook, 1981* (Seoul: EPB, 1981), pp. 303, 407, 469.
34. Ibid., pp. 303, 407, 485; Bank of Korea, *Economic Statistics Yearbook, 1980* (Seoul: BOK, 1980), pp. 58–59.

1974 did continue the rapid rise that had begun in 1970, and the GDP increased nearly 8 percent.[35]

On the eve of the second oil crisis, domestic inflationary pressures in Korea were strong, although perhaps not as strong as they had been in 1974. In 1978 GDP increased 10.6 percent, wholesale prices rose 11.7 percent, and the money supply grew 25 percent. The competitive position of Korean exports was not as strong as it had been in 1974, so the oil price increase could not be passed on as easily to foreign consumers.

The government decided that it must try to counter at least part of the combined effect of domestic inflationary pressures and higher oil prices. It therefore adopted more restrictive monetary and fiscal policies in 1979. Interest rates were pushed up, the annual rate of increase in the money supply was reduced to 20.6 percent, and the surplus of current revenue over current expenditure was increased sharply.[36] The results of these actions were mixed. Wholesale prices rose 18.8 percent in 1979, and the rate of growth of GDP dropped to about 7.0 percent. In 1980 growth of the money supply was even less (at 16.3 percent), GNP at constant market prices declined by slightly more than 6 percent, and inflation continued to grow as wholesale prices rose by 38.9 percent. The country slipped into a recession as unemployment reached 5.2 percent, the highest it had been since 1967. Yet the government was optimistic about an economic recovery, relying mainly on improvement in the trade balance. Export credits were increased to improve Korea's competitiveness overseas.[37] The turnaround of the economy in 1981 was impressive. Real GNP rose 7.1 percent, and wholesale prices increased only 11.8 percent.[38]

Although government policy allowed for the immediate pass-through of oil price increases, energy-use figures registered sizable growth rates after 1974. There are several possible explanations for this increase: (1) the residential sector depends mainly on coal and electricity, the prices

35. EPB, *Korea Statistical Yearbook, 1981*, pp. 243, 469.

36. Ibid., pp. 303, 332–33, 485.

37. Ibid., pp. 69, 303, 407; EPB, *Handbook of Korean Economy, 1979* (Seoul: EPB, 1979), p. 347 (historical interest rates); Bank of Korea, *Monthly Economic Statistics*, vol. 35 (December 1981), p. 132 (for 1980 GNP); Norman Thorpe, "South Korean Five-Year Plan Calls for Slower Growth; Exports Likely to Remain Driving Force of Economy," *Asian Wall Street Journal*, August 3, 1981, p. 14.

38. Korea Economic Institute (Washington, D.C.), *Korea's Economy*, vol. 1 (June 1982), pp. 2, 4.

of which did not increase as much as petroleum products; (2) the cost of gasoline mainly affects people with higher incomes, who can better afford a price increase or can avoid it by using company cars; (3) shifting industrial energy use from oil to coal takes time; (4) businesses have relatively little incentive to reduce energy consumption because they can easily pass on increased costs to the consumer.[39]

The Korean government adopted somewhat different policies in the two oil crises. In 1974 it in effect accepted a large increase in the general price level in the interest of continued rapid economic growth. The depressing effect of the increase in imported oil prices was moderated by passing on much of the increase to purchasers of Korean exports. In 1979, when the competitive position of Korean exports was weaker than it had been in 1974, the government adopted more restrictive fiscal and monetary policies. These policies seem to have reduced the inflationary effect of the rise in oil prices, but at the expense of a lower rate of economic growth.

Taiwan

When the oil crisis of 1973–74 hit, Taiwan's economy was under strong inflationary pressures. The WPI (1971 = 100) rose from 104.5 in 1972 to 128.3 in 1973.[40] Most nominal energy prices, however, remained unchanged in 1973; real energy prices therefore declined sharply.

Faced by a fourfold increase in the price of imported crude oil, the government applied drastic measures to squeeze inflation out of the economy. Prices subject to government control, including energy prices, were increased in early 1974. As a result of these actions the WPI index jumped to 180.4, an increase of 40.6 percent over 1973.[41] Credit was simultaneously tightened. The rediscount rate, which had already been increased 26.5 percent in 1973, was raised another 11.6 percent in 1974. The money supply increased only 7.0 percent in 1974, compared with 49.3 percent in 1973.[42]

---

39. These points were made by Ronald Ridker at a seminar at Resources for the Future, Washington, D.C., December 16, 1981.

40. Directorate-general of Budget, Accounting, and Statistics, Executive Yuan, *Statistical Yearbook of the Republic of China, 1980* (Taipei: Veterans Printing Works, 1980), table 182, pp. 450–51.

41. Council for Economic Planning and Development, *Taiwan Statistical Data Book, 1981* (Taipei: CEPD, 1981), p. 166.

42. Ibid., p. 134.

The predictable—and intended—consequence of these policies was a sharp but brief reduction in the rate of ecnomic growth. Real GDP had increased 12.9 percent in 1973, but it grew only 1.1 percent in 1974 and 4.8 percent in 1975. In 1975 the government relaxed its credit restraints, and in 1976 rapid economic growth resumed. GDP rose 13.7 percent in that year. The temporary drop in economic growth in 1974–75 was the price paid for success in checking inflation. The WPI fell 5.1 percent in 1975 and did not regain the 1974 level until 1977. In 1978, on the eve of the second oil crisis, it was at 187.2, only 3.8 percent above its level in 1974. The balance of trade (on the basis of foreign exchange settlements) had turned negative in 1973. The deficit decreased in 1974, and successively larger surpluses were experienced from 1975 through 1978.[43]

In response to the approximate doubling of imported crude oil prices in 1979, the government appears to have followed policies similar to those applied in 1974. Administered prices, including energy prices, were again increased (but less than in 1974), and credit was tightened. The WPI rose to 213.1 in 1979, an increase of 13.8 percent over 1978. The rediscount rate was increased 33.3 percent, and the money supply grew only 7.0 percent in 1979, compared with 34.1 percent in the previous year.[44]

This strategy has not been as effective in checking inflation in Taiwan as it was after the first oil crisis, and it has not had as severe an impact on economic growth. The WPI rose another 21.5 percent in 1980 to a new high of 259.7, but the rate of growth of GDP dropped only from 13.4 percent in 1978 to 7.8 percent in 1980. It can be argued that this result is satisfactory. The inflationary threat was less in 1979 than it had been in 1973–74, and there was less reason to pay a high price in lost growth to stop inflation.

The government's use of deliberate price increases to curtail consumer purchasing power had two opposite effects on energy prices: nominal energy prices increased along with other administered prices, but the rise in real energy prices was moderated by the increase in the general price level. A rough general index of real energy prices rose 45.9 percent from 1973 to 1975.[45] The overall increase in this index from 1973 to 1979 was 64 percent. Despite this substantial increase in real energy prices,

43. Ibid., pp. 21, 166, 183.
44. Ibid., pp. 134, 166.
45. This index was derived from Directorate-general of Budget, Accounting, and Statistics, *Statistical Yearbook of the Republic of China, 1980*, table 182, pp. 450–51.

energy consumption per capita rose significantly more rapidly than GDP per capita in the period 1973–79. In other words, developments that increased energy intensity were stronger than the depressing effect of higher energy prices on energy consumption.

The 1973–74 oil crisis caused the governments of both South Korea and Taiwan to take a strong interest in diversifying energy sources as a means of reducing dependence on imported oil. Both countries have adopted ambitious nuclear energy programs, and both are increasing their imports of steam coal. Neither government moved very rapidly to encourage energy conservation, but, perhaps prodded by the second oil crisis of 1979, both now have well-developed conservation programs, especially for industry. The 1979–80 crisis also appears to have caused both governments to pass increases in international oil prices on to consumers more rapidly.

Taipei has given higher priority to checking inflation than has Seoul. This difference was most sharply revealed at the time of the first oil crisis, when Taipei precipitated a severe, if brief, recession to contain inflationary pressures. Seoul, however, accepted a large increase in prices rather than curtail rapid economic growth. The policies of the two governments in response to the second oil crisis were more similar. Both adopted moderately restrictive policies that struck a compromise between the goals of stability and growth.

CHAPTER FIVE

Energy Security
and Foreign Policy

NORTHEAST ASIA is an important factor in the international oil market, and it is becoming increasingly so in the markets for coal, liquefied natural gas, and uranium. More is involved, however, than simply trade. Japan and, to a lesser extent, South Korea and Taiwan have also invested in efforts to find and develop foreign energy resources. Moreover, energy considerations have strongly influenced the foreign policies of all three countries.

Volume, Nature, and Source of Northeast Asia's Energy Imports

All three Northeast Asian countries are substantial importers of oil and coal, and Japan also imports liquefied natural gas. Table 5-1 shows the amounts of these fuels imported by each of the three countries in 1965, 1973, and 1979, and table 5-2 expresses these imports as percentages of total world trade in each of the fuels. In 1965 Northeast Asia had not yet begun to import gas, but its imports of both oil and coal were about one-tenth of total world imports of those fuels. In 1973 Northeast Asia accounted for one-sixth of world imports of oil and one-third of world imports of coal, but only 3 percent of world imports of gas. In 1979 Northeast Asia's share of oil imports was still about one-sixth, but its share of coal imports had fallen to about 30 percent. Its share of gas imports, however, had risen to about one-tenth.

Table 5-1. *Northeast Asian Imports of Oil, Coal, and Gas in 1965, 1973, and 1979*
Ten billions of kilocalories[a]

Country and year	Oil	Coal	Gas	Total
Japan				
1965	81,636	11,782	0	93,418
1973	271,548	40,739	2,569	314,856
1979	277,740	40,154	18,523	336,417
South Korea				
1965	1,476	− 63	0	1,413
1973	13,313	290	0	13,603
1979	29,425	2,745	0	32,170
Taiwan				
1965	1,616	− 3	0	1,613
1973	9,457	93	0	9,550
1979	18,395	1,415	0	19,810
Northeast Asia				
1965	84,728	11,716	0	96,444
1973	294,318	41,122	2,569	338,009
1979	325,560	44,314	18,523	388,397

Sources: Japan: Institute of Energy Economics, *Energy Balance Tables* (Tokyo: IEE, various years); Korea (see table 3-2): Ho Tak Kim, "Korea's Energy Experience in the '70s and Short- to Medium-Term Options Open to Korea" (Seoul: Korea Energy Research Institute, September 1980, draft), Economic Planning Board, *Major Statistics of Korean Economy, 1980* (Seoul: EPB, 1980), and Korea Energy Research Institute, *Basic Data for Energy Policy Study* (Seoul: KERI, 1980); Taiwan: Ministry of Economic Affairs, Energy Committee, *Taiwan Energy Statistics, 1980* (Taipei: MEA, 1980). Numbers are rounded.
   a. Ten billion kilocalories approximately equal a thousand metric tons of oil equivalent (t.o.e).

Table 5-2. *Northeast Asia's Share of World Imports of Oil, Coal, and Gas in 1965, 1973, and 1979*
Percent

Item	Japan			South Korea			Taiwan			Northeast Asia total		
	1965	1973	1979	1965	1973	1979	1965	1973	1979	1965	1973	1979
Oil	10.89	15.35	14.08	0.22	0.97	1.47	0.21	0.55	0.95	11.32	16.87	16.50
Coal	10.65	32.91	26.29	0.07	0.35	2.71	0.00	0.03	0.53	10.72	33.29	29.53
Gas	0.00	3.01	10.39	0.00	0.00	0.00	0.00	0.00	0.00	0.00	3.01	10.39

Sources: Oil, 1965 and 1973: U.S. Department of the Interior, Bureau of Mines, *International Petroleum Annual, 1965*, pp. 5–6, and *1973*, p. 9; 1979: World, Japan, Korea—United Nations, *Yearbook of World Energy Statistics, 1979* (New York: UN, 1979), pp. 427, 442–43, Taiwan—Council for Economic Planning and Development, *Taiwan Statistical Data Book, 1980* (Taipei: CEPD, 1980), p. 88. Coal, 1965: UN, *World Energy Supplies, 1950–74* (New York: UN, 1976), pp. 124, 144, 147; 1973, 1979: Institute of Gas Technology, *Energy Statistics*, vol. 4 (second quarter, 1981), pp. 90, 96. Gas, 1973, 1979: UN, *Yearbook of World Energy Statistics, 1979*, pp. 836, 844. Numbers are rounded.

Table 5-3. *Sources of Imports of Oil, Coal, and Gas by Japan, South Korea, and Taiwan in 1979*
Percent[a]

Exporting country	Japan			South Korea		Taiwan	
	Oil	Coal	Gas	Oil	Coal	Oil	Coal
Australia	. . .	42.7	. . .	. . .	35.7	. . .	54.1
Brunei	. . .	. . .	32.4	. . .	. . .	5.9	. . .
Canada	. . .	17.2	. . .	. . .	16.3	. . .	2.6
China	3.0	1.9	. . .	. . .	0.8	. . .	. . .
Indonesia	14.7	. . .	54.5	. . .	. . .	3.0	. . .
Iran	11.8	. . .	. . .	10.7	. . .	. . .	. . .
Iraq	5.5	. . .	. . .	. . .	. . .	*	. . .
Kuwait	9.3	. . .	. . .	32.3	. . .	46.2	. . .
Saudi Arabia	31.1	. . .	. . .	53.3	. . .	39.3	. . .
South Africa	. . .	3.2	. . .	. . .	. . .	. . .	16.3
United Arab Emirates	10.5	. . .	7.4	. . .	. . .	4.5	. . .
United States	. . .	28.7	5.8	. . .	24.2	. . .	22.4
USSR	. . .	3.5	. . .	. . .	. . .	. . .	. . .
Vietnam	. . .	*	. . .	. . .	4.2	. . .	. . .
Other	14.1	2.8	. . .	3.8	18.8	1.1	4.6
Total	100.0	100.0	100.0	100.0	100.0	100.0	100.0

Sources: Japan: Ministry of Finance, *Japan Exports and Imports* (Tokyo: MOF, December 1979), pp. 119–30; for sources of oil and coal imported from China, see fn. 1 of this chapter. Korea: Office of Customs Administration, *Statistical Yearbook of Foreign Trade* (Seoul: OCA, December 1979), pp. 89–115. Taiwan: Inspectorate General of Customs, *Monthly Statistics of Trade* (Taipei: IGC, December 1979), pp. 4-106 to 4-130.
* Less than 3.0 percent; the amount reflected in the percentage has been assigned to "other."
a. Numbers shown are percentages of the total value of imports of a given fuel; numbers are rounded.

Japan was responsible for all of the gas imports and most of the oil and coal imports. The shares of Korea and Taiwan in Northeast Asia's oil and coal imports, however, have increased substantially. In 1979 Korea accounted for 9.0 percent of the region's oil imports and 6.2 percent of its coal imports. Taiwan's shares of the region's oil and coal imports in 1979 were 5.7 percent and 3.2 percent, respectively.

Table 5-3 gives the percentages of oil, coal, and gas imported by each of the Northeast Asian countries in 1979 from various exporting countries. This table illustrates both the heavy dependence of these countries on oil from the Persian Gulf and the reduction in that dependence that has been achieved by substituting coal or gas for oil. Shifting to coal or gas has meant, however, reliance on only a few major suppliers. The dominant position of Australia as an exporter of coal to Northeast Asia stands out, although the United States and Canada are also important. Japan, the only Northeast Asian importer of liquefied natural gas, relies mostly on Indonesia and Brunei for that fuel.

Nearby Asian areas (China and the Soviet Union) are not important sources of energy for the Northeast Asian countries, although other Pacific Basin countries (Australia, United States, Brunei, Indonesia) are. As table 5-3 shows, in 1979 Japan obtained only 3.0 percent of its oil imports and 1.9 percent of its coal imports from China.[1] In the same year, only 3.5 percent of Japan's coal imports came from the USSR.

In the late 1970s South Korea began to import significant quantities of coal from China (0.8 percent of total coal imports by value in 1979). Arrangements for coal shipments from China to South Korea were made through brokers in Hong Kong, but the coal moved directly from Chinese to South Korean ports. The Chinese reportedly suspended coal shipments to South Korea in 1982 in response to protests by North Korea. Taiwan does not trade directly with mainland China or with the USSR and is not known to import any energy materials from those areas indirectly.

The Northeast Asian countries import all of the uranium used as fuel in their nuclear power plants. On the basis of nuclear generating capacity, it is possible to estimate that in 1979 the Northeast Asian countries needed approximately the following amounts of natural uranium[2] in metric tons: Japan, 2,130;[3] South Korea, 83; Taiwan, 181; Northeast Asia total, 2,394. These estimates of requirements are subject to a considerable range of error. If average capacity factors were below the assumed 70 percent, the estimates may be too high. If many reactors received their initial fuel loads (which contain more uranium than replacement loads) in 1979, the estimates may be too low. Actual imports

1. In 1979 Japan imported 8.0 million metric tons of oil and 1.5 million metric tons of coal from China. The oil figure is from "Lagging Production Trims China's Oil Exports," *Oil and Gas Journal*, vol. 79 (March 23, 1981), p. 58; the coal figure is from U.S. Department of Energy, *Interim Report of the Interagency Coal Export Task Force*, draft for public comment, DOE/FE-0012 (Government Printing Office, January 1981), p. 57.

2. A 1,000-MW(e) (megawatt [electric]) light water reactor is assumed to require 4,260 metric tons of natural uranium over thirty years of operation at 70 percent of capacity and using low-enriched fuel with a tails assay of 0.2 percent ^{235}U. See International Nuclear Fuel Cycle Evaluation, *Fuel and Heavy Water Availability*, Report of INFCE Working Group 1 (Vienna: International Atomic Energy Agency, 1980), p. 66. In 1979 Japan had a nuclear generating capacity of 15,000 MW(e), Korea 587 MW(e), and Taiwan 1,272 MW(e).

3. Japanese uranium requirements in 1979 have been estimated at 2,580 metric tons. See Thomas L. Neff and Henry D. Jacoby, *The International Uranium Market*, MIT Energy Laboratory Report no. MIT-EL 80-014 (Cambridge, Mass.: MIT, December 1980), p. 5-8.

of uranium would also be above or below requirements if stocks were being built up or drawn down.

Total world production of uranium outside Communist areas was 33,900 tons in 1978. If the estimates of requirements above are realistic, about 7 percent of world production went to meet requirements in Northeast Asia.[4] Japan's major sources of uranium are Canada, South Africa (including Namibia), and Australia. Korea gets its uranium from Canada and the United States. Taiwan is largely dependent on the United States, although it has bought some uranium from South Africa and will probably buy more in the future.

Detailed data on the uranium imports of Korea and Taiwan are not available. Japan's total uranium import commitments in 1979, however, have been estimated at 7,740 metric tons, broken down as follows (in percentages): Australia, 9.6; Canada, 32.3; France, 11.6; Niger, 5.2; South Africa, 12.3; Rio Tinto Zinc, 25.2; United States, 3.9.[5] Some of the uranium supplied by Rio Tinto Zinc may come from Canada and Australia, but a large part may be from Namibia.[6] Japan appears to be pursuing a policy of building up its uranium stocks by importing more uranium each year than is consumed.

All three Northeast Asian countries have all or most of their uranium enriched by the U.S. Department of Energy. Japan obtains some enrichment services from the French-controlled Eurodif (Société Européenne d'Usine de Diffusion Gazeuse).

Efforts to Increase the Reliability of Energy Imports

Japan has long pursued a policy of trying to diversify sources of imported energy and to increase the percentage of energy imports controlled by Japanese firms.[7] This policy has been most fully developed for oil, but it has also been applied to other fuels.

4. Organization for Economic Cooperation and Development, OECD Nuclear Energy Agency and International Atomic Energy Agency, *Uranium: Resources, Production and Demand* (Paris: OECD, December 1979), p. 22.
5. Neff and Jacoby, *International Uranium Market,* p. 5-5. Numbers are rounded.
6. Ibid., p. 5-4.
7. For a review of Japan's efforts to increase the reliability of its oil imports in the years immediately following the 1973–74 oil crisis, see Yuan-li Wu, *Japan's Search for Oil: A Case Study on Economic Nationalism and International Security* (Stanford, Calif.: Hoover Institution Press, 1977).

Table 5-4. *Overseas Projects Assisted by the Japan National Oil Company as of December 1980*

	Oil production		Oil exploration	
Location	Number	Capital (billions of yen)	Number	Capital (billions of yen)
Middle East	5	148.2	3	8.7
Southeast Asia	5	54.9	14	93.9
Africa	3	36.3	1	0.5
North America	0	0.0	5	14.5[a]
South America	1	9.7	2	1.8
Total	14	249.1	25	119.4[a]

Source: Petroleum Association of Japan, *Petroleum Industry in Japan, 1980* (Tokyo: PAJ, 1981), pp. 10–13. Numbers are rounded.
a. Plus $2.86 million.

The Japanese government provides several incentives for Japanese firms to engage in oil exploration and development in other countries: tax concessions; insurance against losses from war, expropriation, and the like; and loans from the Export-Import Bank of Japan and the Japan National Oil Company (JNOC). The JNOC was formed in 1978 as the successor to the Japan Petroleum Development Corporation (JPDC), which had been established in 1967.[8] Like the JPDC, the JNOC's principal function is to provide financial assistance to oil exploration and production projects. It also extends financial support to companies maintaining oil stockpiles.

The JNOC is permitted to finance 80 percent of domestic oil projects and 70 percent of overseas projects (80 percent if the principal company involved is Japanese). The JNOC can also guarantee 60 percent of loans made by other financial institutions to overseas oil projects. As of December 1980, the JNOC was assisting fourteen overseas oil production projects and twenty-five overseas oil exploration projects (table 5-4). The heavy emphasis on the Middle East for production projects and on Southeast Asia for exploration projects contrasts with the relatively light involvement of the JNOC (and the Japanese firms that it assists) in other areas.

During the period 1961–77, ten Japanese companies supplied 9.8

8. Information on the JNOC is from a descriptive brochure (undated) published by the company, supplemented by conversations with JNOC officials.

percent of Japan's crude oil imports. Eight of these companies, which accounted for 9.2 percent of total oil imports, had been financially assisted by the JNOC. One company, the Arabian Oil Company (one of those aided by the JNOC), overshadowed the rest and contributed 7.2 percent of total oil imports.[9]

The effort to obtain an increased and reliable flow of steam coal from overseas is coordinated by the Japan Coal Development Company (JCDC), which was established by the Japanese electric utility industry (nine private companies and one public) in January 1980. JCDC engages in a wide range of activities: prospecting, investing in coal mines (sometimes joint ventures), contracting for the purchase of coal, reselling coal to its sponsoring utility companies, and establishing coal stockpiles and distribution centers. Its principal areas of operation are Australia, Canada, the United States, and South Africa. In Australia it operates through a locally incorporated subsidiary.[10]

An idea of the scope of Japan's overseas energy activities may be gained from table 5-5 (at end of chapter), in which only a partial inventory of Japan's recent energy-related investments is given. Some of the investments listed are still in the discussion stage, and not all investments turn out well—for example, construction of a large petrochemical plant at Bandar Khomeini in Iran[11] and a joint U.S.-Japanese-German solvent-refined coal project in the United States (SRC-II).[12] Nevertheless, the extent of Japanese energy investments overseas is impressive.

On a more modest scale, Korea and Taiwan have also engaged in energy-related activities overseas. Korean firms are involved in coal mining projects in Australia, Indonesia, and the Philippines and in uranium prospecting in the Sudan, Gabon, and Paraguay.[13] (The prospecting activity in Paraguay is a joint effort of Korea Electric Power

9. Japan Petroleum Consultants, *Japan Petroleum and Energy Yearbook, 1978* (Tokyo: JPC, 1978), p. F-18.

10. Japan Coal Development Co., *Company Profile* (Tokyo: JCDC, March 1980), and conversations with JCDC officials (Tokyo, November 1980).

11. On May 17, 1983, Mitsui and Company and Iran's National Petrochemical Company agreed to resume construction, with the Iranian company agreeing to pay all additional costs involved. The agreement clearly cannot take effect until the end of the Iran-Iraq war. See *Japan Economic Journal,* May 24, 1983, p. 1.

12. Changes of heart in Washington and Bonn seem to have ended the joint effort; *New York Times,* June 25, 1981.

13. *Korea Newsreview,* February 21, 1981, p. 17.

Company, Taiwan Power Company, and a U.S. firm.) Korea Development Company has joined Pertamina, the state oil company of Indonesia, in exploring for oil off the coast of Java.[14]

The Korean Oil Development Corporation (KODC), a state-owned corporation, was created in 1978. KODC has joined Petroleum National Derhab Company, a state-owned Malaysian corporation, to build a petrochemical complex in Malaysia costing $800 million. When completed in 1986, the complex is expected to produce yearly 150,000 metric tons of low-density polyethylene, 40,000 tons of polypropylene, and 50,000 tons of ethylene glycol.[15]

Private Korean construction companies have become heavily involved in overseas work. In 1980 they obtained contracts valued at $8.26 billion. Some of these contracts are in energy-exporting countries, although many of them are not directly related to the production or processing of energy materials. In mid-1981 four South Korean construction companies were engaged in projects in Libya, and 100,000 Koreans were at work on construction projects in Saudi Arabia.[16]

In Taiwan, ensuring future imports of steam coal and uranium has been the responsibility of the Taiwan Power Company, and ensuring future imports of oil (and, eventually, of gas) has been the responsibility of the Chinese Petroleum Company. Both state-owned companies have worked in most cases by entering into long-term purchase agreements. The Chinese Petroleum Company, however, has participated in joint ventures to explore for oil in the Philippines, Colombia, and the United Arab Emirates. The Overseas Petroleum Investment Corporation, a subsidiary of the Chinese Petroleum Company, has signed an agreement with Pertamina and Amoco Indonesia Petroleum Company to explore for oil in Indonesia. In 1981 the government of the Marshall Islands awarded the Overseas Petroleum Investment Company the exclusive right of oil exploration both on land and offshore.[17] In addition to the exploratory effort in Paraguay already mentioned, the Taiwan Power Company recently entered into a joint venture to explore for uranium in Gabon.[18]

14. "International Briefs," *Oil and Gas Journal,* February 9, 1981, p. 54.

15. *Korea Newsreview,* June 6, 1981, p. 12.

16. *Asian Wall Street Journal Weekly,* September 21, 1981, pp. 1, 20.

17. Chinese Petroleum Corporation, *Annual Report, 1981* (Taipei: China Color Printing Co., 1982), p. 4.

18. Foreign Broadcast Information Service, *Worldwide Report: Nuclear Development and Proliferation,* JPRS 78695, no. 108 (August 6, 1981), p. 10.

Construction firms from Taiwan have been active in some of the oil-exporting countries, and Taiwan has made energy-related foreign investments. The government of the Republic of China has signed an agreement with Saudi Arabia to build a $357 million fertilizer plant in Saudi Arabia that will produce 500,000 tons of urea annually from cheap natural gas. Both construction costs and output will be shared equally. In a somewhat similar arrangement, Chang Chun Petrochemical Company will invest $10 million in the construction of a methanol plant in Indonesia and will be given an option to buy 100,000 tons of the plant's annual output of 340,000 tons.[19]

Energy and the Foreign Policies of the Northeast Asian Countries

The various specific efforts to increase the reliability of energy imports that have been described are of course manifestations of the foreign policies of the governments behind such efforts. The goal of energy security also affects the foreign policies of those governments in more general ways and must be balanced against other objectives.

The adjustments in Japanese foreign policy during and after the 1973–74 Arab oil embargo illustrate the interaction between energy and other interests. As noted in chapter 2, Japan's need for Arab oil caused it to lean toward the Arab side in the Arab-Israeli dispute. Japan did not, however, break diplomatic relations with Israel, and in November 1973 it stopped short of an unqualified endorsement of the Arab position. Japan was undoubtedly inhibited by a desire not to offend unduly the United States, with which it has important economic and security ties.[20] A few months later, Japan responded positively to the U.S. initiative that led to the creation of the International Energy Agency, but it made

19. Phil Kurata, "Searching for New Friends Abroad," *Far Eastern Economic Review*, vol. 109 (August 8, 1980), pp. 44–45.

20. On November 22, 1973, the Japanese government issued a formal statement that called on Israel to withdraw from all Arab territories occupied in the 1967 war and expressed support of "the rights of the Palestinian people for self-determination." But the statement also supported the legitimacy of the state of Israel by expressing "respect for the integrity and security of the territory of all countries in the area." See *Japan Times*, November 23, 1973, p. 1.

clear its strong desire to avoid confrontation between the oil-importing and oil-exporting countries.[21]

To the extent that Japan's interest in good relations with the United States conflicts with its interest in good relations with major oil-exporting countries, Japan's policy is to try to find some middle ground. Japan has thus expanded its investments in the Middle East, but it also appears to have responded to U.S. concerns by moderating its efforts to negotiate explicit barter agreements for oil. Similarly, Japan has adopted a low posture toward successive increases in the price of oil by the Organization of Petroleum Exporting Countries but was responsive to U.S. complaints during the 1979–80 oil crisis that purchases by Japanese firms on the spot market were unnecessarily driving up the price of oil.

Japan's energy relations with China and the USSR provide an interesting contrast. Japan's energy dealings with China have not been greatly complicated by nonenergy considerations. This has not been true of Japanese-Soviet energy relations.

Importing oil and coal from China is quite consistent with Japan's general policy of expanding trade and improving relations with that country. At least since the early 1970s, Japan has had no reason to fear that energy imports from China would adversely affect U.S.-Japanese relations. The only problems encountered in the energy trade with China have been secondary ones, such as price, the physical qualities of Chinese oil and coal, resource availability, and the advantages and disadvantages of shifting some of Japan's import business from traditional suppliers to China.

From the narrow view of energy supply, Japan has a strong interest in gaining access to the large oil, gas, and coal resources of Soviet Siberia. Pursuit of this interest, however, has been complicated by other factors. Relations between Japan and the USSR are cool at best, in part because of the longstanding distrust of the Russians by the Japanese (in marked contrast with their more benevolent view of the Chinese). Negative Japanese attitudes toward the USSR also can be traced to fears

---

21. At the Washington Energy Conference in February 1974, Japanese Foreign Minister Ohira declared in part: "My delegation is participating in this Conference of major oil-consuming countries, anticipating that it will be the first step in building a harmonious relationship between the oil-producing and consuming countries. . . . Japan feels it is of primary importance to realize, as early as possible, a constructive dialogue with the oil-producing countries." See U.S. Department of State, *Washington Energy Conference*, Document 9, February 11, 1974 (unofficial translation), p. 8.

aroused by increasing Soviet military strength and resentment over continued Soviet occupation of several small islands (the "Northern Territories") seized by the USSR at the end of World War II. The efforts of the United States to check Soviet expansion are also a constraint on Japanese economic relations with the USSR, a constraint that no longer exists in Japanese relations with China.[22]

The differences in the policy contexts and environments in which Japan must pursue its energy interests with respect to China and the USSR by no means ensure that China will be more important than the USSR as a source of energy supply. Other factors, including resource availability, are also important.

In pursuing their energy interests, South Korea and Taiwan are subject to greater constraints than Japan. Both are involved in a bitter struggle with strong domestic Communist adversaries. The merits and feasibility of many possible international moves, including some in energy, may depend on how they affect this domestic political struggle. Both South Korea and Taiwan are also sensitive to the possibility that their actions may affect their relations with the United States.[23]

Some of the energy-related activities of South Korea and Taiwan are designed to counter actual or anticipated moves of their Communist national rivals as well as to increase the reliability of energy supplies. For example, Libya is not now a significant source of energy for South Korea, but the extensive activity of South Korean construction firms in Libya serves both as a counterweight to North Korean activity and as a means of earning needed foreign exchange.[24] The recent decision of the Taiwan Power Company to enter into a joint uranium exploration venture in Gabon can be justified as an effort to ensure energy supplies, but

22. As part of sanctions against the USSR, the U.S. government blocked delivery of U.S.-made equipment needed in a joint Japanese-Soviet project to explore for oil and gas off the island of Sakhalin. See *Japan Economic Journal*, February 23, 1982, p. 1, and May 11, 1982, p. 6. The partial lifting of sanctions in November 1982 appeared to open the way to exporting the equipment in question. See *Japan Economic Journal*, November 23, 1982, p. 1.

23. Despite its lack of diplomatic relations with the United States and the termination of the formal U.S. security commitment, Taiwan may be more vulnerable to further adverse changes in U.S. policy—and, therefore, more sensitive to the effect of its actions on the remaining U.S. connection—than is South Korea.

24. An indication of the scale of South Korean activity in Libya is provided by the inauguration of regular flight service by Korean Airlines between Seoul and Tripoli. See *Korea Newsreview*, September 5, 1981, p. 4.

it also serves to strengthen the political position of the Republic of China in Francophone Africa.

The struggles of South Korea and Taiwan with their domestic Communist adversaries also have the effect of ruling out some energy options, or of making them less attractive. Neither South Korea nor Taiwan imports energy from their Communist national rivals. South Korea, as noted earlier in this chapter, has imported some coal from China, but this trade will not be resumed if Peking decides that such trade is too damaging to its ties with Pyongyang. For both political and economic reasons, South Korea would probably like to import energy materials from Siberia. The Soviets, however, have thus far been even more careful than the Chinese not to offend North Korean sensibilities. The USSR might be interested in trading with Taiwan, but such trade would be ruled out by the firm anti-Communist policy of the government of the Republic of China.

Taiwan operates under some handicaps that do not affect South Korea. The lack of diplomatic relations with most nations rules out some energy arrangements and makes others more difficult. For example, the possibility of Taiwan's importing a Canadian deuterium-uranium (CANDU) nuclear power reactor was barred by the Canadian government on political grounds. Taiwan has been remarkably successful in maintaining economic ties with countries that do not recognize it diplomatically. Taiwan continues to import fuels from countries with which it does not have formal ties.[25]

Nuclear energy has involved all three Northeast Asian countries in difficulties with the United States.[26] Disagreements between Japan and the United States have centered on Japan's operation of a pilot reprocessing plant and on Japan's shipments of spent nuclear fuel to reprocessing plants in France and England. South Korea and the United States have clashed over the Korean desire to buy a reprocessing plant from a French firm, and Taiwan and the United States have had differences over the operation of a nuclear research laboratory on Taiwan. The positions taken by the United States prevailed in the

25. That Taiwan gets over half its coal imports from Australia is a case in point. (See also chapter 9, table 9-3.)

26. These difficulties are discussed in greater detail in Joseph A. Yager, ed., *Nonproliferation and U.S Foreign Policy* (Brookings Institution, 1980), pp. 27–30, 45, 79–80.

disputes with Korea and Taiwan. Temporary compromises were worked out between the United States and Japan, and eventual agreement on a permanent solution appears likely.

In summary, the importance of Northeast Asia in international energy markets has greatly increased since the mid-1960s. In 1979 one-sixth of all internationally traded oil outside Communist countries went to Northeast Asia. In the same year, Northeast Asia accounted for nearly 30 percent of all coal imports and over 10 percent of all natural gas imports. Something under one-tenth of world uranium production outside Communist nations went to meeting the requirements of Northeast Asia's nuclear power plants.

The countries of Northeast Asia remain heavily dependent on oil from the countries of the Persian Gulf. That dependence, however, has been somewhat reduced by shifting to use of other fuels. Australia is the dominant supplier of coal to all three Northeast Asian countries. Most of the gas imported by Japan comes from Indonesia and Brunei. Canada appears to be the major supplier of uranium to Northeast Asia.

The heavy import dependence of the Northeast Asian countries has inevitably affected their foreign policies. Japan and, to a lesser extent, South Korea and Taiwan have made extensive efforts to increase the reliability of energy imports by investing in efforts to find and develop energy resources in other countries. Other investments in energy-exporting countries—some of them energy related—have been designed to solidify relations with those countries.

The pursuit of the goal of energy security has also had more general effects on the foreign policies of the Northeast Asian countries. Balancing that goal against other important objectives, including maintaining good relations with the United States, has been a continuing problem.

Table 5-5. *Recent Japanese Foreign Energy Investments*

Country of investment	Type of investment and parties involved	Description
Australia	Uranium mine development: Kansai Kyushu, Shikoku Electric Power Companies	Industrial Bank of Japan and several Western banks to finance mine with $390 million loan; each power company is to import 100 short tons of uranium yearly starting in 1982
Victoria State	Brown coal liquefaction project: Nippon Brown Coal Liquefaction Co. (Kobe Steel Ltd., Mitsubishi Chemical Industries Ltd., Nissho Iwai Corp., Idemitsu Koan Co., Asia Oil Co.)	Each company owns 20 percent of Nippon Brown Coal Liquefaction's 500 million yen capital; Nippon is to run brown coal liquefaction project supplied with coal from Victoria State
	Nuclear enrichment project proposal by Australia, also to involve Japan and United States	To be completed in late 1980s with annual capacity of 1,000-ton separative work units or the energy requirement for 9 million kilowatt nuclear power plants
	Steaming coal supply agreement: Howard Smith, Coal & Allied Industries, Ube Industries	Two Australian companies to supply $1.5 billion of steaming coal over twenty-year period; first-year price $33 per ton, to be renegotiated in subsequent years
	Steaming coal supply agreement: Workworth Associates (consortium of Australian and Japanese companies), Electric Power Development Company of Japan	Electric Power Development Co. to import 5 million tons of steaming coal from northern New South Wales mines
Queensland	Coking coal supply deal: Mitsui and Co. (with six steel producers), Thiess Dampier Mitsui Coal Ltd. (of which Mitsui owns 20 percent equity share)	3.3 million tons of coking coal to be purchased yearly for fourteen and one-half years starting in 1983
Canada Alberta Province	Coal mining: Mitsui and Co. (with six steel producers), Gregg River Coal Ltd. (Canada)	Japanese to purchase 40 percent mining rights of Gregg River Coal by paying development costs equivalent to value of rights ($182 million in Canadian dollars); Japan to import 31 million tons over fifteen and one-half years.

Northeast British Columbia	Metallurgical coal project: Quintette Coal Ltd. (affiliate of Denison Mines Ltd.)	Japan to pay for infrastructure and capital costs for development of project (construction costs estimated at $700 million); Japan to receive 75 million metric tons over fifteen years starting in 1983
Beaufort Sea	Oil exploration: Japan National Oil Co. (JNOC), Dome Petroleum Ltd.	Cost to Japan: $400 million
Fort McMurray, Alberta	Tar sands project: Japan Oil Sands Co. (with JNOC), Petro-Canada	Development of a tar sands retrieval project
British Columbia	Petrochemical plants: 5-company consortium—Mitsubishi Chemical Industries Ltd., Mitsubishi Petrochemical Co., Mitsubishi Corp., Occidental Petroleum Corp. (U.S.), Dome Petroleum Ltd. (Canada)	Construction of petrochemical plants to be completed in 1985 (total project cost, $2 billion); Japan to import petrochemical products made from liquefied natural gas (LNG)
China	Japanese $1.3 billion credit offer	$1 billion in government loans to purchase Japanese goods and $300 million provided by commercial banks; funds in part to complete petrochemical plant in Daqing
Gulf of Bohai	Oil exploration: Chengbei Oil Development Corp. (with JNOC), China Petroleum Corp.	Japanese to finance 0.5–1.0 billion yen yearly for three to five years; Japan to receive 42.5 percent share by late 1982 for fifteen years.
Sitaigou, Shanxi Province	Coal project: Mitsui Mining Co., Coal Industry Ministry (China)	Japan to lend 1 billion yen to be repaid with coal; production target, 5 million tons a year
Senkaku Islands	Oil exploration: China, Japan, United States	China's proposal of offshore exploration and development around Senkaku Islands, which both Japan and China claim
Tatung	Coal mine development: Mitsui Mining Co., Coal Industry Ministry (China)	To be financed by Export-Import (Ex-Im) Bank (Japan); estimated yield, 4 million metric tons a year

Table 5-5 *(continued)*

Country of investment	Type of investment and parties involved	Description
China *(continued)* Various provinces	Oil development: China, Japan	China asks Japan aid in developing four oil fields: Talim Basin in Xianjiang Uygur Autonomous Region (est. reserve: 5 billion barrels), Chaidamu Basin in Qinghai Province (5.3 billion barrels), Sichuan Basin in Sichuan Province (3.1 billion barrels), Huabei Basin in Hobei and Honan Provinces (6.8 billion barrels); JNOC to finance $500 million for Huabei Basin out of the $2 billion the Ex-Im Bank of Japan is to provide for Chinese energy development projects
Egypt Gulf of Suez	Oil development: Egyptian Petroleum Development Co. (Tokyo)	Third Japanese firm successfully to develop oil commercially (the other two successful firms are Arabian Oil Co. and Abu Dhabi Oil Co.), will take 30 percent of oil produced; of the remaining 70 percent Egypt will receive 85 percent and Japan 15 percent
Gabon	Oil development: Mitsubishi Petroleum Development Co., ELF Aquitaine (France)	Crude oil commercially produced off Gabon since June 1980, first successful oil production for Mitsubishi (estimated yield by 1981: 7–8 thousand barrels a day); Gabon government to receive 25 percent, ELF and Mitsubishi to split remainder
Indonesia Badak Field, Kalimantan	LNG plant: four Japanese companies (Chubu Electric, Kansai Electric, Osaka Gas, Toko Gas), Pertamina (Indonesia)	To extend LNG plant ($900 million, twenty-year contract to export 3.2 million tons of LNG a year starting in 1983); Chubu to receive 1.5 million tons a year, Kansai Electric 0.8 million, Osaka Gas 0.4 million, Toko Gas 0.5 million
	Oil project: Consortium of twenty-six Japanese companies, Pertamina	Exploration and development of crude oil; Japanese investment to be repaid with crude
	LNG imports: Japanese government proposal	Japan's proposal to lend $500 million to increase oil refining capacity at Cilaciap (central Java) and Balikpapen (east Kalimantan) in exchange for increased LNG imports

Iran Bandar Khomeini	Petrochemical plant: five Japanese companies (Mitsui and Co., Mitsui Toatsu Chemical, Toyo Soda Manufacturing Co., Petrochemical Industries, Japan Synthetic Rubber Co.)	Work has been halted on Japan-Iran petrochemical plant project (Japanese government approved withdrawal from this $3.17 billion venture; Mitsui could ask for up to $434 million from its insurance coverage)
Japan-Korea Continental Shelf	Oil exploration: Nippon Oil Co., South Korean government	Unsuccessful first oil well cost $2 billion; two more wells to be drilled (costs will be equally split between Japan and Korea)
Mexico	Oil development: consortium of thirty-six Japanese companies, Pemex (Mexico)	Japan to extend $500 million credit for oil development in Mexico
	Oil imports: Keidanren, Siderurgica Mexicana S.A., Nacional Financiera S.A.	Japan to help finance a cast and forged steel plant, 11.6 billion yen, and a large diameter steel pipe plant, 10.44 billion yen; in exchange, Mexico to increase crude oil exports to Japan
Peru	Pipeline: Japan	Japan to extend $330 million credit for construction of trans-Andean pipeline; Japan to start receiving 15,000 barrels a day of crude oil in 1981
Saudi Arabia	Oil development: Japanese consortium (led by Mitsubishi), Saudi Arabia Basic Industries (merged with Dow Chemical Enterprise)	Japanese investment is 52.5 billion yen, of which government to contribute 25 billion; Mitsubishi to begin receiving 130,000 barrels a day in 1986
Thailand	Natural gas project: Southeast Asia Petroleum Exploration Co., Mitsui Oil Exploration Co., Union Oil Co. of California	Two natural gas wells in Bay of Thailand (estimated daily output by 1981 or 1982, 290 million cubic feet)

Table 5-5 *(continued)*

Country of investment	Type of investment and parties involved	Description
United States	Solvent-refined coal (SRC-II): Gulf (U.S.), Mitsui (Japan), Ruhrkohle (Germany)	Cost of project $100 million (Japanese and German shares 25 percent each, U.S. share 50 percent); SRC-II canceled (Japan disappointed—considered project symbolic, hoped to use coal liquefaction technology on deposits it is developing in Australia and China)
Beluga, Alaska	Sub-bituminous coal development: Electric Power Development Co. of Japan, Japan Coal Development Co., Tokyo Electric Power Co., Placer AMEX Inc. (U.S.), BHW (U.S.)	Sub-bituminous coal development project (estimated exports to Japanese electric companies by 1984–85, 4 million tons a year)
Southern United States	Oil and gas exploration: Japan Petroleum Exploration Co., Calgary (Alberta subsidiary of Bow Valley Exploration Ltd.)	Total cost $10 million (equally divided)
USSR Siberia	Coal mine: Ex-Im Bank of Japan, Soviet government	Ex-Im Bank to lend $949 million ($40 million for coal mine project, $909 million for forestry resource development)
Okhotsk Sea (north of Sakhalin Island)	Oil development: Sakhalin Oil Development Cooperation Co. (Japanese consortium 70 percent owned by Japan Petroleum Development Corp., with one foreign major, Gulf Oil)	SODECO expenditures on exploration and development to be repaid with crude
Yakutsk (eastern)	Natural gas exploration: U.S. and Japanese banks	Bank of America and Japanese banks to provide $25 million each for exploration for natural gas (estimated yield for export, 353 billion cubic feet a year for each country)

Sources: Australia: Japanese Economic Journal, Sept. 23, 1980, p. 7, Aug. 5, 1980, p. 7, June 17, 1980, p. 7; New York Times, Dec. 26, 1979, p. D3. Canada: Japan Economic Journal, Nov. 11, 1980, p. 1, Feb. 17, 1981, p. 5; Wall Street Journal, Feb. 3, 1981, p. 6; Oil and Gas Journal, Jan. 26, 1981, p. 98. China: Washington Post, Sept. 8, 1981, p. A1; Japan Economic Journal, May 26, 1981, p. 4, June 30, 1981, p. 16; Oil and Gas Journal Newsletter, Nov. 3, 1980; Japan Economic Journal, Apr. 29, 1980, p. 1, Sept. 9, 1980, p. 6. Egypt: Japan Economic Journal, Aug. 5, 1980, p. 7. Gabon: Japan Economic Journal, July 22, 1980, p. 6. Indonesia: Japan Times, Apr. 15, 1981, p. 5; Oil and Gas Journal, Jan. 26, 1981, p. 98; Japan Economic Journal, Nov. 11, 1980, p. 5. Iran: New York Times, Sept. 3, 1981, p. D4. Japan-Korea: Japan Economic Journal, July 15, 1980, p. 6. Mexico: Oil and Gas Journal, Jan. 26, 1981, p. 98; Japan Economic Journal, Sept. 9, 1980, p. 6. Peru: Oil and Gas Journal, Jan. 26, 1981, p. 98. Saudi Arabia: Japan Economic Journal, Mar. 10, 1981, p. 16. Thailand: Japan Economic Journal, Dec. 12, 1980, p. 6. United States: New York Times, June 25, 1981, p. D1; Japan Economic Journal, Sept. 33, 1980, p. 6, Oct. 14, 1980, p. 6. USSR: New York Times, June 11, 1981, p. D7; Economist, Oct. 22, 1977, p. 93; Wall Street Journal, Apr. 1, 1976, p. 12; Oil and Gas Journal, Apr. 26, 1976, p. 91.

CHAPTER SIX

Future Final Energy Requirements in Japan

JAPAN's future energy requirements and the way in which Japan seeks to fill those requirements may have major consequences, both for international energy markets and for Japan's relations with other nations. In this chapter the question of future energy requirements by final consumers is considered. The next chapter estimates how these energy requirements might be met, first, in terms of various fuels and, second, in terms of the geographical origins of those fuels.

The detailed analysis of Japanese energy requirements and supplies that follows will assume, with few variations, present trends in international economic and political conditions and no significant change in Japan's own energy policies. These assumptions will be relaxed later in the study to deal both with several international contingencies and with hypothetical changes in Japanese energy policies.

Future Energy Consumption: The Central Case

The future consumption of energy in Japan or any other country depends upon a wide variety of economic, political, and social influences, many of which cannot be quantified. Prediction is clearly impossible. The best that can be done is to project possible future consumption on the basis of what seem to be reasonable assumptions about a few of the major factors involved. The result will be not a view of the future, but only a framework for thinking about future possibilities.

81

The approach that will be followed here is to set forth central, or reference, case projections for each of the major energy-consuming sectors and to show how such projections might be altered by changing key assumptions. The 1980s will be considered in greater detail than the 1990s.

Central case projections for all major energy-consuming sectors will assume, or be consistent with, a 5.0 percent average annual growth rate of gross domestic product and fairly stable energy prices in real terms.[1] These assumptions are in no sense predictions. The Japanese economy may grow at an average annual rate above or below 5.0 percent.[2] In all probability real energy prices will not remain stable, but assuming that they will serves a useful analytical purpose. The effects of possible price changes can then be considered in alternatives to the central case.

The Industrial Sector

In recent years manufacturing has accounted for well over 90 percent of industrial energy consumption. In FY1980 the breakdown of energy consumption in this sector was as follows:[3]

1. A flat assumption of no change in real energy prices was rejected on practical grounds. Such an assumption would have interfered with using economic relationships from the recent past (1976–79) to project future energy consumption because those relationships unavoidably reflect small changes in real energy prices. Such an assumption would also have precluded using the best available projections of energy consumption in manufacturing because these assume increases in nominal crude oil prices that could cause small increases in real energy prices. (See the next section of this chapter.)

2. The Japanese Ministry of International Trade and Industry has declared that Japan has the potential of achieving 5 percent annual growth until 1990. See *Japan Economic Journal*, October 12, 1982, p. 5. Earlier in 1982, however, the Economic Council, which advises the prime minister, issued a report entitled "Japan in 2000" that stated that Japan could achieve an annual growth rate of 4 percent in the 1980s and 1990s. See *Japan Economic Journal*, June 22, 1982, p. 1. On the basis of this report, the Japanese cabinet adopted a program in August 1983 that called for an average annual rate of growth of 4 percent through fiscal year 1990. See *Japan Economic Journal*, August 16, 1983, p. 1. Achieving an average annual growth of 5.0 percent for the period 1979–85 now appears unlikely. Real gross national product, GNP, grew 5.3 percent in FY1979, 4.5 percent in FY1980, and 3.3 percent in both FY1981 and FY1982. See Economic Planning Agency, *Monthly Economic Report* (Tokyo: EPA, September 1983), p. 7. Unless otherwise indicated, all references to years in this chapter are to Japanese fiscal years, which begin on April 1 of the year specified.

3. From energy balance table prepared by Institute of Energy Economics (Japan), *Energy Balances in Japan (1980)* (Tokyo: IEE, 1981), pp. 60–61. Figures are rounded.

	Percentage
Manufacturing	92.5
Agriculture and forestry	2.5
Fishing	3.6
Mining (excluding fuels)	0.4
Construction	0.9

In estimating future energy consumption in the industrial sector, attention will be concentrated on manufacturing. It will be assumed that the energy consumption of other components of the industrial sector will continue to represent about 7.5 percent of total sectoral consumption.

A forecast of energy consumption in Japanese manufacturing industry in 1990 prepared by the Institute of Energy Economics (Japan) (IEE) will be adopted as the reference case for that year. IEE forecasts of energy consumption and manufacturing production are set forth in table 6-1. Actual data for 1973 and 1979 are also presented for purposes of comparison. The IEE forecasts assume that Japan's GNP will grow at an average annual rate of 5.0 percent during the 1980s and that the nominal international price of crude oil will be $90 a barrel in 1990. Other assumptions are given in the note to table 6-1.

Real energy prices paid by Japanese consumers would rise much less than the assumed 10.9 percent annual increase in the nominal price of crude oil from 1980 to 1990. The cost of crude oil is only a fraction of the total cost of producing and delivering refined petroleum products to consumers (other major costs are refining, transportation, and taxes), and an increase in the price of crude oil would have only an indirect (and smaller) effect on the prices of other fuels. Moreover, the rise in nominal energy prices within Japan brought about by the assumed increase in international crude oil prices would have to be deflated by assumed changes in the general price level to determine the increase in real energy prices.[4] It appears that the assumed increase in nominal crude oil prices would cause real energy prices to rise very little.[5]

On the stated assumptions, the IEE projections show energy consumption in manufacturing increasing at an average annual rate of 2.1 percent from 1979 to 1990, in contrast with an average annual decrease

4. An assumption concerning the exchange rate between the U.S. dollar and the yen would also be required because international oil prices are expressed in dollars.

5. This is also the view of analysts at the IEE, where the projections set forth in table 6-1 were prepared.

Table 6-1. *Energy Consumption in Japanese Manufacturing Industries per Physical Unit of Output, 1973, 1979, and Estimates for 1990*[a]

Item[b]	1973	1979	Annual rate of change, 1973–79 (percent)	1990	Annual rate of change, 1979–90 (percent)[c]
Paper and pulp					
A. Energy consumption	6,534.0	5,833.0	−2.0	8,456.0	3.4
B. Production index	122.3	134.0	1.5	218.9	4.6
A/B	53.4	43.5	−3.4	38.6	−1.1
Chemicals					
A. Energy consumption	39,129.0	36,878.0	−1.0	48,404.0	2.5
B. Production index	119.9	135.8	2.1	220.0	4.5
A/B	326.4	271.6	−3.0	220.0	−1.9
Cement					
A. Energy consumption	8,771.0	8,506.0	−0.5	9,367.0	0.9
B. Output (thousands of metric tons)	78,700.0	88,600.0	2.0	116,300.0	2.5
A/B	0.1115	0.0960	−2.5	0.0806	−1.6
Other ceramics					
A. Energy consumption	5,339.0	5,648.0	0.9	6,941.0	1.9
B. Production index	129.3	131.1	0.2	193.1	3.6
A/B	41.3	43.1	0.7	35.9	−1.6
Iron and steel					
A. Energy consumption	57,187.0	49,938.0	−2.2	59,214.0	1.6
B. Output (thousands of metric tons)	120,000.0	113,000.0	−1.0	130,600.0	1.3
A/B	0.4765	0.4419	−1.3	0.4533	0.2
Nonferrous metals					
A. Energy consumption	4,628.0	4,302.0	−1.2	5,173.0	1.7
B. Production index	130.7	145.8	1.8	209.7	3.4
A/B	35.4	29.5	−3.0	24.7	−1.6
Food and tobacco					
A. Energy consumption	4,871.0	4,728.0	−0.5	5,720.9	1.7
B. Production index	98.9	112.4	2.1	144.9	2.3
A/B	49.2	42.1	−2.6	39.5	−0.6
Textiles					
A. Energy consumption	6,586.0	5,515.0	−3.0	6,011.0	0.8
B. Production index	117.7	106.9	−1.6	139.4	2.4
A/B	55.9	51.6	−1.3	43.1	−1.6
Other					
A. Energy consumption	17,956.0	17,449.0	−0.5	25,268.0	3.4
B. Production index	121.8	148.3	3.3	277.1	5.8
A/B	147.4	117.7	−3.7	91.2	−2.3
Total manufacturing					
A. Energy consumption	151,000.0	138,795.0	−1.4	174,554.0	2.1
B. Production index	118.4	136.4	2.4	228.4	4.8
A/B	1,275.3	1,017.6	−3.7	764.2	−2.6

Source: Institute of Energy Economics (Japan), paper (in Japanese) presented at IEE energy symposium, Tokyo, December 1981; calculations prepared in May 1981. Numbers are rounded.

a. Unless otherwise indicated, all references to years in this and subsequent tables in this chapter are to Japanese fiscal years, which begin on April 1 of the year specified.

b. Figures for energy consumption are in ten billions of kilocalories (approximately thousands of metric tons of oil equivalent, t.o.e.).

c. The following average annual growth rates were assumed for the period 1980–90: GNP, 5.0; private final consumption, 4.8; public final consumption, 2.0; private household investment, 6.0; private facility investment, 5.2; public investment, 4.5; exports, 6.2; imports, 6.0; wholesale price index, 4.0. In addition, it was assumed that the consumer price index would increase at an average annual rate of 6.5 percent from 1980 to 1985 and 5.5 percent from 1985 to 1990. The price of oil was assumed to rise to $90 a barrel in 1990 (presumably, in current U.S. dollars).

Table 6-2. *Share of Energy-intensive Japanese Manufacturing Industries in Sectoral Energy Consumption, 1973, 1979, and Estimates for 1990*
Percent

Industry	1973	1979	1990
Paper and pulp	4.3	4.2	4.8
Chemicals	25.9	26.5	27.7
Ceramics, stone, and clay products	9.3	10.2	9.3
Primary metals	40.9	39.0	36.9
Other manufacturing	19.5	20.0	21.2
Total	100.0	100.0	100.0

Source: Same as table 6-1. Numbers are rounded.

of 1.4 percent from 1973 to 1979. As is shown in table 6-2, four energy-intensive industries are expected to continue to account for a large part of energy consumed by the manufacturing sector. The anticipated decline in the share of primary metals in total manufacturing energy consumption is the most significant feature of the figures in the table. Energy consumption in primary metals is projected to be only 4.2 percent greater in 1990 than it was in 1973. Total energy consumption in manufacturing is projected to increase 15.6 percent over the same period.

The recent decline in the intensity of energy use in manufacturing is projected to continue during the 1980s, but at a slower rate. Measured by the ratio of energy consumption to the index of industrial production, energy intensity fell at an average annual rate of 3.7 percent from 1973 to 1979, and it is projected to fall at an average annual rate of 2.6 percent from 1979 to 1990. If energy intensity is measured by the ratio of energy consumption to value added, the rate of decline for 1973–79 is 6.3 percent, that for 1979–90, 4.0 percent (table 6-3).

The difference in the rates of change in the two measurements of energy intensity reflects an increase in the value added per unit of physical output in manufacturing. In 1973 each unit in the manufacturing production index was associated with 394 billion yen of value added.[6] This figure rose to 468 billion yen in 1979, and, on the basis of the projections in tables 6-1 and 6-3, it will increase to 551 billion yen in 1990. The last two figures suggest a possible leveling off of the increase in value added per unit of output.

6. All yen figures in this chapter are constant 1975 yen.

Table 6-3. *Energy Consumption in Japanese Manufacturing Industries per Unit of Value Added, 1973, 1979, and Estimates for 1990*

Item[a]	1973	1979	Annual rate of change, 1973–79 (percent)	1990	Annual rate of change, 1979–90 (percent)
Paper and pulp					
A. Energy consumption	6.5	5.8	−2.0	8.4	3.4
B. Value added	1.5	1.9	4.0	2.4	2.1
A/B	43.3	30.5	−5.7	35.0	1.3
Chemicals					
A. Energy consumption	39.1	36.9	−1.0	48.4	2.5
B. Value added	3.5	5.4	7.5	10.1	5.9
A/B	111.7	68.3	7.9	47.9	−3.2
Ceramics, stone, and clay products					
A. Energy consumption	14.1	14.2	0.1	16.3	1.3
B. Value added	2.4	1.9	−3.7	2.4	2.1
A/B	58.8	74.7	4.1	67.9	−0.9
Primary metals					
A. Energy consumption	61.8	54.2	−2.2	64.4	1.6
B. Value added	5.3	7.2	5.2	9.3	2.4
A/B	116.6	75.3	−7.0	69.2	−0.8
Subtotal (energy-intensive industries)					
A. Energy consumption	121.5	111.1	−1.5	137.5	2.0
B. Value added	12.7	16.4	4.4	24.2	3.6
A/B	95.7	67.7	−5.6	56.8	−1.6
Other					
A. Energy consumption	29.5	27.7	−1.0	37.1	2.7
B. Value added	34.0	47.4	5.7	101.6	7.2
A/B	8.7	5.8	−6.5	3.6	−4.2
Total manufacturing					
A. Energy consumption	151.0	138.8	−1.4	174.6	2.1
B. Value added	46.7	63.8	5.3	125.8	6.4
A/B	32.3	21.8	−6.3	13.9	−4.0

Sources: For energy consumed in manufacturing, IEE (unpublished calculations, May 1981). For value added in 1973, 1979, and 1985 (not shown in table), Japan Economic Research Center, *Five-Year Economic Forecast, 1981–85* (Tokyo: JERC, March 1981), pp. 26–27. Value added in 1990 was estimated on the assumption that GDP would increase 5 percent annually from 1985 to 1990 and that value added in manufacturing would be 40 percent of GDP in 1990. The shares of various industries in value added in manufacturing were calculated on the assumption that their relative rates of growth from 1985 to 1990 would be the same as those the Japan Economic Research Center had projected from 1980 to 1985. Numbers are rounded.

a. Figures for energy consumed in manufacturing are in ten trillions of kilocalories (approximately millions of t.o.e.). Figures for value added in manufacturing are in trillions of 1975 yen. The ratio of energy consumption to value added (A/B) is in kilocalories per yen.

Not surprisingly, the fall (both actual and projected) in the intensity of energy use in manufacturing is accompanied by a decrease in the importance of the most energy-intensive industries—paper and pulp; chemicals; ceramics, stone, and clay products; and primary metals. In 1973 those industries contributed 27.2 percent of total value added in manufacturing (in 1975 yen). In 1979 they contributed 25.7 percent, and in 1990 they are projected to contribute 19.2 percent. Most of the decline in the importance of this group of industries is attributable to the ceramics (including cement) and primary metals industries. The share of the chemicals industry in total value added in manufacturing actually increases slightly.

The ratio of energy consumed in manufacturing to GDP fell from 1973 to 1979. If the GDP increases at an average annual rate of 5.0 percent in the 1980s (as is assumed in table 6-1), this decline is projected to continue in the period 1979–90. Past and projected future values of this ratio in selected years (in kilocalories per yen) are: 1973, 10.34; 1979, 7.56; 1990 (estimated), 5.55.[7] These figures indicate that less than three-quarters as much energy consumption in manufacturing was required to produce a unit of GDP in 1979 as was required in 1973. By 1990 the ratio of energy consumed in manufacturing to GDP is projected to be only a little more than half as large as it was in 1973.

A decrease in the intensity of energy use was largely responsible for the fall in this ratio between 1973 and 1979. Over three-fourths of the projected decline is attributable to a fall in the intensity of energy use in manufacturing industries as measured by the ratio of energy consumption to value added. The remainder of the decline would result from a decrease in the importance of energy-intensive industries in the manufacturing sector. That decrease would more than overcome the effect of an increase in the share of manufacturing in the total economy.

Estimating energy consumption in manufacturing industries in the 1990s is, if anything, a more speculative endeavor than doing so for the 1980s. Extending the reference case to the year 2000 is, however, a useful way of bringing out the relationships among several important

7. Sources are: GDP for calendar years 1973 and 1979, table 2-1; GDP for FY1979 and FY1985, Japan Economic Research Center, *Five-Year Economic Forecast, 1981–85* (Tokyo: JERC, March 1981), pp. 26–27. GDP for FY1990 is estimated assuming a 5.0 percent annual growth rate. Figures for energy consumed in manufacturing are from IEE, *Energy Balance Tables* (Tokyo: IEE, October 1980), and projections by IEE prepared in May 1981.

Table 6-4. *Energy Consumption in Japanese Manufacturing Industries and Related Economic Variables, 1973, 1979, and Estimates for 1990, 1995, and 2000*

Item	1973	1979	1990	1995	2000
GDP (trillions of 1975 yen)	146.9	185.4	314.4	401.2	511.9
Value added in manufacturing (trillions of 1975 yen)	46.7	63.8	125.8	160.5	204.8
Index of manufacturing production (1975 = 100)	118.4	136.4	228.4	282.6	350.1
Energy consumed in manufacturing (ten trillions of kilocalories)[a]	151.0	138.8	174.6	190.9	209.0

Sources: For GDP in 1973, 1979, and 1985 (not shown in table), Japan Economic Research Center, *Five-Year Economic Forecast, 1981–85*, pp. 26–27; 5.0 percent annual growth rate assumed for years after 1985. For value added, index of manufacturing production, and energy consumption in 1973, 1979, and 1990, see tables 6-1 and 6-3; value added in 1995 and 2000 is estimated under assumptions stated in the text. Numbers are rounded.

a. Ten trillion kilocalories approximately equal a million t.o.e.

variables. The following assumptions, which are linked to the projections for 1990 in tables 6-1 and 6-3, will be used in projecting manufacturing energy consumption in 1995 and 2000:

—Real GDP will continue to increase by 5.0 percent annually (the rate of increase assumed for the 1980s).

—Value added in manufacturing will stabilize at 40 percent of GDP (the share assumed for 1990).

—The ratio of value added in manufacturing to the index of manufacturing production will increase by 3.0 percent every five years.

—The ratio of energy consumed in manufacturing to the index of manufacturing production will decrease by 2.5 percent annually (slightly below the projected 2.6 percent annual rate of decrease for the period 1979–90).

—Real energy prices will not change.

On the basis of these assumptions, table 6-4 presents projections of GDP, value added in manufacturing, the index of manufacturing production, and energy consumed in manufacturing for 1995 and 2000. Figures for 1973, 1979, and 1990 are also shown. Under these projections, the ratio of energy consumed in manufacturing to GDP continues to fall. In 1995 the ratio is 4.76 kilocalories per yen, and in 2000 it is 4.08 kilocalories per yen.

The projections in table 6-4 also imply changes in the income elasticity of demand for energy used in manufacturing. Projected annual rates of increase in energy consumed in manufacturing and in real GDP for various future periods are shown in table 6-5. The ratio between such

Table 6-5. *Projected Average Income Elasticities of Energy Demand in Japanese Manufacturing, 1979–2000*

Years	Annual rate of increase in energy consumed in manufacturing (percent) (1)	Annual rate of increase in GDP (percent) (2)	Average income elasticity of energy demand (3)[a]
1979–90	2.1	4.7	0.45
1990–95	1.8	5.0	0.36
1995–2000	1.8	5.0	0.36

Source: Figures in columns 1 and 2 are based on table 6-4. Numbers are rounded.
a. Column 3 = column 1/column 2. Because energy consumed in manufacturing declined from 1973 to 1979, the ratio for that period is negative—hence meaningless as a measurement of income elasticity of energy demand—and is not included.

rates of increase in a given period is the average income elasticity of energy demand for that period. The assumptions concerning GDP and energy prices will be varied later in this chapter as a means of constructing high and low alternatives to the reference case for all sectors over the entire period 1979–2000. At this point, it is useful to note the effects of varying the relationships between variables that were assumed in projecting the reference case for energy consumed in manufacturing in 1995 and 2000. Several illustrative cases in which only one assumption is changed may be cited:

—If the share of value added in manufacturing declined, say, to 37.5 percent of GDP in 1995 and 35.0 percent in 2000, rather than remaining at 40 percent, energy consumed in manufacturing would be 6.3 percent lower in 1995 and 12.5 percent lower in 2000 than in the reference case.

—If the trend toward increased value added per physical unit of production ends (that is, if the ratio of value added to the production index is the same in 1995 and 2000 as it was projected to be in 1990), energy consumed in manufacturing would be 3.0 percent higher in 1995 and 6.2 percent higher in 2000 than in the reference case presented in table 6-4.

—If the trend toward increased efficiency of energy use ends (that is, if the ratio of energy consumption to the production index is the same in 1995 and 2000 as it was projected to be in 1990), energy consumption in manufacturing would be 13.1 percent higher in 1995 and 28.0 percent higher in 2000 than in the reference case.

It is difficult to judge whether the reference case is more plausible than the above variations or any of the innumerable other variations that might be constructed. The leveling off of the share of manufacturing in Japanese GDP assumed in the central case, however, is consistent with the experience of several other industrialized countries. A continued rise in value added per physical unit of manufacturing output also appears reasonable given Japan's shift toward increasingly advanced production technologies. The rise in value added and continued improvements in the efficiency of energy use probably go hand in hand because high-technology processes tend not to be energy intensive. In an energy study such as this one, however, the assumption that the intensity of energy use in manufacturing will continue to fall requires closer examination.

At first sight, the past and possible future declines in the importance of the four most energy-intensive industrial categories (paper and pulp; chemicals; ceramics, stone, and clay products; and primary metals) would appear to favor this assumption. The share of this group of industries in total value added in manufacturing fell from 27.2 percent in 1973 to 25.7 percent in 1979 and has been projected to decline to 21.1 percent in 1985.[8] If the estimated growth rates of energy-intensive and other manufacturing industries from 1979 to 1985 are projected, the share of the energy-intensive industries will fall still further: to 19.2 percent in 1990, 17.0 percent in 1995, and 14.9 percent in 2000.

Trends in value added and in energy consumption, however, do not coincide. As pointed out earlier in this section, the share of the four energy-intensive industries in total energy consumed in manufacturing actually rose slightly from 1973 to 1979 and has been projected to continue its gradual increase in the 1980s. The declining importance of the energy-intensive industries in value added has undoubtedly moderated the rise in the share of these industries in energy consumed in manufacturing. The fundamental question, however, is what will happen to the ratio of energy consumption to physical output in these industries.

Energy consumption per unit of physical output can be reduced by (1) operating existing facilities more efficiently, (2) investing in improvements in existing facilities, (3) building new, more energy-efficient facilities to replace existing facilities or to expand capacity, and (4) importing intermediate products that embody energy. Most of the potential reductions from the first of these methods were probably

8. Japan Economic Research Center, *Five-Year Economic Forecast*, pp. 26–27.

achieved in the years immediately following the 1973–74 oil crisis. Investing in improvements, and to an even greater extent investing in new facilities, depends on profit margins that have been narrowed in some industries by rising energy costs. Investing in new facilities is, of course, also much more likely in industries whose markets are expanding.[9] Importing intermediates is an attractive energy-saving alternative for firms that cannot finance expensive improvements or the construction of new facilities.

The efficiency of energy use is affected in the short run by the percentage of an industry's capacity that is being used. If production processes and the structure of the industry permitted the closing of entire plants or production lines, the least efficient plants or lines would tend to be closed in slack periods, and energy consumption per unit of output would fall. If entire plants or production lines were operated substantially below capacity, however, efficiency would suffer, and energy consumption per unit of output would tend to rise.[10] An examination of several energy-intensive industries is instructive.

IRON AND STEEL. Energy consumption per metric ton of crude steel fell 7.3 percent from 1973 to 1979, but it is projected in table 6-1 to increase 2.6 percent from 1979 to 1990. Energy consumption per ton would therefore be 4.9 percent below the 1973 level in 1990.

These projections contrast with more optimistic projections made by the IEE in 1978.[11] These earlier projections estimated that energy consumption per ton of crude steel would be 11.1 percent below the 1973 level in 1990. An important reason for the difference between the 1978 and 1981 projections is a lowering of estimates of future steel production. In 1978 steel production was expected to reach 160 million metric tons in 1990. Three years later, in 1981, steel production was estimated to be

9. The Industrial Bank of Japan points out that Japan's industrial plants are steadily growing older. Over the past ten years (1971–81), the average age of all industrial productive facilities increased 0.52 years annually. In FY1981, the average age reached 7.32 years. The problem is most serious in the basic material-related industries. The principal plants of the petrochemical industry were over ten years old in 1981. See *Japan Economic Journal*, June 30, 1981, p. 1.

10. Cartels to dispose of excess capacity in certain "structurally depressed" industries are exempt from the antimonopoly laws. Some of these industries, including aluminum smelting, paper, and petrochemicals, are energy intensive. See *Japan Economic Journal*, February 22, 1983, p. 4.

11. IEE, "Present State and Future Potentials of Energy Conservation in Japan (Summary)," supplement to *Energy in Japan*, no. 45 (June 1979), p. 6.

130.6 million tons in 1990.[12] Lower production means that fewer energy-efficient new plants will be built.[13] It also means that a larger percentage of existing facilities will probably operate below capacity and in a relatively inefficient manner.[14]

CHEMICALS. Energy intensity in the chemicals industry (measured by the amount of energy consumption associated with every unit of the production index) fell 16.8 percent from 1973 to 1979 and is projected in table 6-1 to fall another 19.1 percent from 1979 to 1990. Analyzing energy consumption in this industry is made difficult by the lack of data on the fuels (in Japan, largely naphtha and other petroleum products) used as feedstocks in the production of petrochemical products. From one point of view this nonenergy use of energy materials should be set aside in calculating changes in the energy efficiency of the chemicals industry. From another point of view—and the one followed here—chemical feedstocks are taken from fuel supplies and therefore must be regarded as part of energy consumption.

Considerable room appears to exist for improving the energy efficiency of at least the petrochemical component of the chemicals industry. Energy efficiency of production processes can be improved by shifting to products with higher value added and by replacing existing plants with larger, more modern ones.[15] Moreover, intermediate petrochemicals can be imported from countries in which fuels used as feedstock (largely natural gas) are cheaper than the petroleum products used in Japan.

In 1976 the twenty-nine Japanese plants making ethylene—the primary intermediate petrochemical product—required on average 10.2 million kilocalories of energy (including energy material used as feedstock) to make 1 metric ton of product. If smaller plants were scrapped, seven new plants built, and the total number of plants reduced to nineteen

12. The reduced estimate for 1990 may prove to be high. In October 1982, the executive vice-president of NIPPON Steel Corporation estimated that Japanese production of crude steel in FY1984 would be only 105 million metric tons.

13. A change in the structure of output of the steel industry currently under way may somewhat lower the energy consumption per metric ton of steel that is projected in table 6-1. The share of high-quality steel made from scrap in electric furnaces is increasing. This rising share increases the use of electricity but decreases the use of coking coal to make pig iron. (Conversation at the IEE, Tokyo, February 1982.)

14. In 1978 the Japanese steel industry had a total capacity of 166.8 million metric tons. See Robert W. Crandall, *The U.S. Steel Industry in Recurrent Crisis: Policy Options in a Competitive World* (Brookings Institution, 1981), p. 26.

15. See Takeshi Hijikata, "Petrochemical Industry: Crisis and Opportunities," *Keidanren Review*, no. 75 (June 1982), pp. 2–5.

(but with a 16.5 percent increase in total capacity), the specific energy consumption rate might be reduced by an estimated 18.3 percent to 8.33 million kilocalories per ton.[16] Further replacement of smaller old plants and an additional increase in total capacity might lower the specific energy consumption rate still further.

Scrapping old plants and building larger new ones will be financially attractive only if sales are expanding. But the Japanese petrochemical firms—in particular, the producers of ethylene—are suffering from foreign competition. Firms in this industry are much more likely to increase their imports of intermediate products than to invest in more energy-efficient facilities.[17] The value (in current yen) of Japan's imports of organic chemicals, which include petrochemical intermediate products, rose from 185.4 billion yen in 1976 to 381.5 billion yen in 1980, an increase of 106 percent.[18] Imports of ethylene dichloride increased 289 percent in 1978 alone.[19]

The projections of energy efficiency in the chemicals industry in table 6-1 do not take full account of the effect of possible future increases in imports of intermediate petrochemical products.[20] One reason to expect an increase in such imports is the interest of Japanese petrochemical firms in investing in productive facilities in overseas areas with cheaper feedstocks.[21]

CEMENT. Energy consumption per metric ton of product in the cement industry fell 13.9 percent from 1973 to 1979. A further decline of 16.0 percent from 1979 to 1990 is projected in table 6-1. Energy consumption per ton is projected to be 27.7 percent below the 1973 level in 1990.

16. IEE, "Present State and Future Potentials," p. 10.

17. See Katsuhiko Suetsugu, "Where Will Basic Industries Go? Entrepreneurs Now Must Reform Their Set-up for International Business," *Japan Economic Journal,* October 13, 1981, p. 24. In mid-1983 the Japanese government was reported to be working on plans to encourage reduction of the capacities of some components of the petrochemical industry. See *Japan Economic Journal,* June 7, 1983, p. 1.

18. Prime Minister's Office, Statistics Bureau, *Japan Statistical Yearbook, 1981,* p. 315.

19. Japan External Trade Organization, *White Paper on International Trade, Japan 1979 (Summary)* (Tokyo: JETRO, 1979), p. 107.

20. Conversation at the IEE, Tokyo, August 1981.

21. See *Japan Economic Journal* on Canada, November 11, 1980, p. 1, and September 29, 1981, p. 9; on Saudi Arabia, March 10, 1981, p. 16; on Indonesia, September 8, 1981, p. 16; and on Alaska, September 29, 1981, p. 9. A more recent article in the *Japan Economic Journal* reaffirmed the likelihood of Japanese investments in overseas plants to produce basic and intermediate chemicals but reported that many such investments were being delayed because of doubts concerning their profitability in the near term.

The increase in energy efficiency can probably be attributed to the adoption of NSP (new suspension preheater) kilns. The switch to such kilns was virtually required by the imposition of stricter standards concerning nitrate emissions.

ALUMINUM. The aluminum industry, which is a major component of the nonferrous metals industry, increased its energy efficiency 3.1 percent a year from 1973 to 1979 and is projected to continue to improve its energy efficiency at an annual rate of 1.6 percent from 1979 to 1990. These gains are expected to be realized by closing less energy-efficient facilities and by adopting energy-saving technology in facilities that remain in operation.

The most important development under way with respect to the energy intensity of the aluminum industry is not the gradual decline in the amount of energy required to produce a metric ton of aluminum in Japan, but the shift from domestically produced aluminum ingots to ingots imported from producers in the United States and Canada, producers that have the advantage of access to cheaper electric power. Production of aluminum in Japan fell from 1,193 thousand metric tons in 1977, to 1,014 thousand tons in 1979, and to 1,096 thousand tons in 1980.[22] Ingot output resumed its decline in 1981 and 1982. By September 1982, output at an annual rate was slightly less than 300,000 tons.[23] Imports of aluminum and aluminum alloys increased in value from 153 billion yen in 1977 to 407 billion yen in 1981.[24] In physical terms, imports of aluminum ingots were about 690,000 tons in 1979 and 840,000 tons in 1980.[25]

This shift from domestic production to imports is, of course, financially painful for the Japanese aluminum industry. In mid-1981 the cumulative deficits of the six major aluminum smelters were estimated at about 31 billion yen.[26] A survey by the Ministry of International Trade and Industry found that Japanese aluminum ingot prices in mid-1981 were about 100,000 yen a ton higher than American and Canadian prices.[27] The price difference is largely attributable to differences in electricity costs.

The Japanese aluminum smelters and the Japanese government have

---

22. Statistics Bureau, *Japan Statistical Yearbook, 1982*, p. 203.
23. *Japan Economic Journal*, November 16, 1982, p. 8.
24. Statistics Bureau, *Japan Statistical Yearbook, 1982*, p. 323.
25. *Japan Economic Journal*, June 30, 1981, p. 6.
26. Ibid.
27. Ibid., July 28, 1981, p. 5.

reacted in several ways to the loss of international competitiveness. The closing down of excess capacity has already been mentioned.[28] Investments have been made, or are planned, in joint aluminum-producing enterprises in other countries.[29] Special efforts are being made to develop energy saving technology,[30] and the government may also adopt various measures to help the industry.[31]

PAPER AND PULP. The ratio of energy consumption to the production index in the paper and pulp industry fell 18.5 percent from 1973 to 1979. In table 6-1 this ratio is projected to be 11.3 percent below the 1979 level in 1990. This projection contrasts with the less optimistic estimate, made by the IEE in 1978, that energy consumption per metric ton of output would be only 5.6 percent below the 1976 level in 1990.[32] This changed view reflects a better performance of the industry than expected in the period between the two estimates.[33]

The Residential and Commercial Sector

This sector includes miscellaneous energy consumption not logically assignable to any other sector, as well as energy consumption by households and commercial enterprises. In 1979 households accounted for 50.4 percent of sectoral consumption, commercial enterprises for 38.5 percent, and the miscellaneous category for 11.1 percent.[34]

---

28. For example, Japan's largest aluminum smelter, Sumitomo Aluminum Smelting Company, was operating at 290,000 metric tons of annual capacity at the beginning of 1980. In mid-1981 it was operating only at 190,000 metric tons of annual capacity. See *Japan Economic Journal*, June 23, 1981, p. 5.

29. The government, private banks, trading firms, and aluminum smelters are investing in a smelting enterprise with 320,000 metric tons of annual capacity in Brazil. See ibid., July 21, 1981, p. 6.

30. Three Mitsui group companies have been developing a new method of smelting aluminum that uses coke, rather than electricity, as the main energy source. The six major smelters are interested in joining this effort. See ibid., July 21, 1981, p. 6.

31. For a detailed analysis of the problems of Japan's aluminum industry, see Takehiko Hayashi, "Japanese Aluminum Industry Struggling for Survival," *Keidanren Review*, no. 79 (February 1983), pp. 2–10.

32. IEE, "Present State and Future Potentials," pp. 14–17.

33. Using a higher proportion of imported pulp would reduce energy consumption per metric ton of paper products produced, but it is not clear what assumption was made in the IEE projection on this score. The percentage of imported pulp to total pulp supply rose from 9.9 percent in 1973 to 18.5 percent in 1980. See Statistics Bureau, *Japan Statistical Yearbook, 1982*, p. 319.

34. Estimates for residential and commercial energy consumption in 1979 are drawn from Energy Conservation Research Center, "National Living Standards and Residential/

Table 6-6. *Residential Energy Consumption and Disposable Income in Japan, 1976, 1979, and Illustrative Projections for 1990*

Year	Energy consumption (EC) (ten billions of kilocalories)[a]	GDP (trillions of 1975 yen)	Disposable income (DI) (trillions of 1975 yen)	DI/GDP	EC/DI (kilocalories per yen)
1976	26,393	155.6	111.4	0.716	2.369
1979	28,300	181.5	121.9	0.671	2.322
Annual rate of change, 1976–79 (percent)	2.35	5.26	3.04	–2.14	–0.67
Case I (assumes 1979 values of DI/GDP and EC/DI)					
1990	48,367	310.4	208.3	0.671	2.322
Case II (assumes 2.1 percent annual decline in DI/GDP, 0.7 percent annual decline in EC/DI)					
1990	35,416	310.4	164.8	0.531	2.149
Case III (assumes 1.0 percent annual decline in DI/GDP and EC/DI)					
1990	38,794	310.4	186.6	0.601	2.079
Case IV (assumes 1.0 percent annual decline in DI/GDP, 1.0 percent annual rise in EC/DI)					
1990	48,348	310.4	186.6	0.601	2.591

Sources: Figures for GDP and disposable income, 1976–79, are from Organization for Economic Cooperation and Development, *National Accounts of OECD Countries, 1962–79,* vol. 2 (Paris: OECD, 1981), pp. 33, 36. Disposable income was deflated using the consumer price index from Prime Minister's Office, Statistics Bureau, *Japan Statistical Yearbook, 1981* (Tokyo: Statistics Bureau, 1981), p. 396. Figures for residential energy consumption in 1976 are from Kenichi Matsui, "Middle, Long-Term Energy Supply-Demand Projection Forecast for Fiscal 1985, 1990," in IEE, *Energy in Japan,* no. 42 (March 1979), p. 46. Figures for residential energy consumption in 1979 are from Energy Conservation Research Center, "National Living Standards and Residential/Commercial Energy Demand," supplement to IEE, *Energy in Japan,* no. 50 (October 1980). Numbers are rounded.

a. Ten billion kilocalories approximately equal a thousand t.o.e.

RESIDENTIAL ENERGY CONSUMPTION. Energy consumption (EC) by households increased at an average annual rate of 2.35 percent from 1976 to 1979. Over the same period GDP increased at an average annual rate of about 6.0 percent, but disposable income (DI) increased only 3.68 percent annually. The intensity of residential energy use, measured by the ratio of energy consumption to disposable income (EC/DI), fell at an average annual rate of 2.22 percent (table 6-6).

These past relationships—between disposable income and GDP,

Commercial Energy Demand," supplement to IEE, *Energy in Japan,* no. 50 (September 1980), pp. 4, 12. Sectoral total is drawn from IEE, *Energy Balances in Japan, 1980.* Miscellaneous energy consumption is the sectoral total in the energy balance table minus residential and commercial consumption.

DI/GDP, and between residential energy consumption and disposable income, EC/DI—can be used to obtain one possible projection of residential energy consumption in 1990.[35] By assuming changes in these relationships, other possible projections can be obtained. Table 6-6 presents several illustrative cases; in all cases real GDP is assumed to increase at an average annual rate of 5.0 percent.

Case I assumes that the values of DI/GDP and EC/DI that existed in 1979 will continue unchanged until 1990. Case II assumes that DI/GDP and EC/DI will continue to decline at the annual rates (2.1 percent and 0.7 percent, respectively) recorded in the period 1976–79. Case III assumes that from 1979 to 1990 both DI/GDP and EC/DI will decline 1.0 percent annually. Case IV assumes that DI/GDP will decline 1.0 percent annually but that EC/DI will rise 1.0 percent annually.

Cases I and IV produce 5.0 percent annual rates of growth in residential energy consumption from 1979 to 1990. Case II produces the lowest rate of growth in residential energy consumption—1.4 percent. Case III results in an intermediate rate of growth—2.9 percent.

Residential energy consumption can be related to changes in population and the number of households, as well as to changes in disposable income. Table 6-7 presents four illustrative projections that are based on the relationships between residential energy consumption (EC) and either population (P) or the number of households (H) in the recent past. These cases are numbered V to VIII to distinguish them from the four cases in table 6-6. The same projections of population and numbers of households are used in all four cases in table 6-7.

Cases V and VI assume that residential energy consumption varies with population, and cases VII and VIII assume that it varies with the number of households. Cases V and VII use the 1979 relationships between residential energy consumption and population or the number of households, and Cases VI and VIII use the 1976–79 trends in those relationships. The projected annual rate of increase in residential energy consumption under these different assumptions varies from 0.7 percent and 1.9 percent in the two cases that use 1979 relationships (cases V and VII, respectively) to 2.3 percent and 2.7 percent in the two cases that carry forward 1976–79 trends (cases VI and VIII, respectively).

The IEE has published two projections of residential energy con-

---

35. Relationships between these variables in the period 1976–79 are used because real energy prices changed very little during those years.

Table 6-7. *Residential Energy Consumption, Population,*
and Number of Households in Japan, 1973, 1976, 1979,
and Illustrative Projections for 1990

Year	Residential energy consumption (EC) (ten billions of kilocalories)[a]	Population (P) (millions)	Households (H) (millions)	EC/P (thousands of kilocalories)	EC/H (thousands of kilocalories)
1973	23,230	109.1	29.7	2,129	7,822
1976	26,393	113.1	31.8	2,334	8,300
1979	28,300	116.1	33.4	2,437	8,473
	Case V (assumes EC varies with P and EC/P remains at 1979 level)				
1990	30,779	126.3	40.7	2,437	7,562
	Case VI (assumes EC varies with P and EC/P increases at 1976–79 annual rate of 1.5 percent)				
1990	36,261	126.3	40.7	2,871	8,909
	Case VII (assumes EC varies with H and EC/H remains at 1979 level)				
1990	34,652	126.3	40.7	2,744	8,514
	Case VIII (assumes EC varies with H and EC/H increases at the 1976–79 annual rate)				
1990	37,826	126.3	40.7	2,995	9,294

Sources: Population figures for 1973, 1976, 1979 are from Statistics Bureau, *Japan Statistical Yearbook, 1980*, p. 13; household figures for 1973 are from *Japan Statistical Yearbook, 1981*, p. 457. Figures for residential energy consumption in 1976 and 1979 are based on table 6-6; 1973 figures are from IEE energy balance tables. Population in 1990 was projected by the Institute of Population Problems, Ministry of Health and Welfare (Tokyo). Household figures for 1976 and 1979 were estimated by interpolation between the reported figures for 1973 and 1980. Household figure for 1990 was calculated from population projection on the assumption that the number of persons per household would continue to decline 0.71 percent annually (as in the period 1975–80). Numbers are rounded.

a. Ten billion kilocalories approximately equal a thousand t.o.e.

sumption in 1990 that are based on a macroeconomic analysis of the Japanese economy.[36] These projections and the related real GDP growth rates (derived) and rates of change in real international crude oil prices (assumed) are shown below:

36. IEE, "National Living Standards," pp. 5–8. The IEE projected energy consumption per household in 1990 at 9,325 thousand kilocalories under high oil prices and 10,045 thousand kilocalories under low oil prices. It also estimated that the number of households would increase at an annual rate of 1.8 percent from 1979 to 1990. The total number of households would therefore increase from 33.4 million in 1979 to 40.7 million in 1990. Multiplying the projections of energy consumption per household by the latter number and converting to metric tons of oil equivalent (t.o.e.) gives estimates of total residential energy consumption in 1990 of 37.9 million t.o.e. in the high price scenario and 40.9 million t.o.e. in the low price scenario.

	Under high oil prices	Under low oil prices
Annual change in real crude oil prices (percent)	2.0	0.0
Annual change in real GDP (percent)	4.0	4.7
Residential energy consumption in 1990 (millions of t.o.e.)	37.9	40.9

Under the high price scenario residential energy consumption would increase at an average annual rate of 2.7 percent. Under the low price scenario the average annual rate of increase would be 3.4 percent.

The differences among the various projections of residential energy consumption described above serve as a useful reminder of the wide range of future developments that are possible. The practical problem, however, is to select a projection to use as part of the central or reference case. Such a projection should at a minimum be consistent with the two assumptions made for all components of the central case: a 5.0 percent annual growth rate for GDP and little or no change in energy prices in real terms. In addition, the central case projection should be plausible.

All of the projections relating residential energy consumption to disposable income assume a 5.0 percent GDP growth rate (table 6-6). They are also at least neutral with respect to energy prices.[37] Two of the four projections (table 6-7) that relate residential energy consumption to population or number of households (cases V and VII) take no account of the effect of GDP growth on energy consumption and can therefore be excluded from further consideration. The other two projections (cases VI and VIII) implicitly carry forward the trends of the period 1973–79, during which GDP increased 6.0 percent annually and domestic energy prices did not change very much. These projections can be adjusted to make them roughly consistent with the GDP assumption by reducing them by one-sixth. Case VI then projects a 1.9 annual increase in residential energy consumption, and case VII foresees a 2.2 percent annual increase.

37. Case II explicitly carries forward the 1976–79 trend in EC/DI, a trend that reflects the lack of substantial change in domestic energy prices during that period. The index of Japanese fuel and energy prices published by the Organization for Economic Cooperation and Development rose 10.3 percent from 1976 to 1979. When this index is deflated by the index of consumer prices from the same source, a decline of 2.7 percent in real fuel and energy prices results. See OECD Economic Surveys: Japan (Paris: OECD, July 1981), p. 16.

The low price scenario of the IEE roughly satisfies the economic growth and energy price assumptions (4.7 percent annual GDP growth, and constant international oil prices in real terms). The IEE's high price scenario, however, is somewhat out of line with both assumptions (4.0 percent annual GDP growth, and 2.0 percent annual rise in real oil prices).

The following projections of annual percentage rates of growth in residential energy consumption appear to satisfy the economic growth and energy price assumptions: cases I and IV, 5.0; IEE low price scenario, 3.4; case III, 2.9; case VIII (adjusted), 2.2; case VI (adjusted), 1.9; case II, 1.4. These projections are, of course, purely illustrative—they do not even fully define the range of possibility. They do, however, show that various rates of increase in residential energy consumption can be consistent with the assumptions concerning economic growth and energy prices adopted for the central case.

In deciding what projection to use in the central case, it is useful to examine trends and prospects in the various components of residential energy consumption.[38] As shown in table 6-8, these components recorded different annual rates of change from 1976 to 1979 (second to last column); the percentage shares of these components in total consumption in 1976 and 1979 are also shown. Grounds exist for believing that most, if not all, of the annual rates of change shown in table 6-8 will be quite different in the 1980s.

On the one hand, the high rates of increase in energy used to heat water and to operate electric lights and appliances are unlikely to continue. Most Japanese households already have water heaters (gas or electric), and the diffusion rates for electric lights and appliances are very high.[39] Moreover, the expected spread of solar water heaters will somewhat reduce the use of gas and electric heaters, and further increases are possible in the energy efficiency of electric appliances.

On the other hand, the sharp decline in the use of energy to heat space that took place from 1976 to 1979 is unlikely to mark the beginning of a

38. See Tsutomu Toichi, Jun Adachi, and Naoyuki Yamamoto, "International Comparison of Residential Energy Use in the Developed Countries," supplement to *Energy in Japan*, no. 53 (June 1981). The judgments in their paper concerning possible future trends in residential energy consumption are in some respects more optimistic than those presented here.

39. Energy Conservation Research Center, "National Living Standards," pp. 9, 15, and table 3-17.

Table 6-8. *Shares of Various Components in Total Japanese Residential Energy Consumption, 1976 and 1979, and Annual and Assumed Future Rates of Change*
Percent

Item	Share in total residential energy consumption		Annual rate of change in energy consumption	
	1976	1979	1976–79	Future
Heating space	39.7	29.7	−7.2	2.0
Cooling space	2.0	1.5	−7.2	2.0
Heating water	27.2	35.6	11.9	5.0
Cooking	15.7	13.8	−1.6	−2.0
Lighting and electric power	15.4	19.4	11.2	5.0
Total	100.0	100.0	2.3	3.1

Sources: 1976 shares are from Matsui, "Middle, Long-term Energy," p. 45; 1979 shares are from Energy Conservation Research Center, "National Living Standards," p. 4. For assumptions governing future rates of change, see the text. Numbers are rounded.

long-term trend. That decline may well reflect the coincidence of two temporary circumstances: warm weather,[40] and a delayed reaction by consumers to the increase in energy prices in 1974 and 1975. A significant increase in energy consumed to heat space would not be surprising in the 1980s. Only about 4.0 percent of Japanese households had central heating in 1979.[41] Most households rely on a portable kerosene stove supplemented by the traditional *kotatsu* (an electric footwarmer covered by a quilt).

Some small increase in the use of energy to heat space would be expected if recent upward trends in population, number of households, number of occupied dwellings, and floor space per occupied dwelling continue.[42] A more substantial increase in energy use for this purpose would be brought about by the spread of central heating and the use of

40. The nationwide average number of degree days (weighted by the number of households in each region) a year in the period 1974–76 was 1,242. In the period 1977–79, the average was 1,109. See IEE, *Declining Energy Demand and Future Trends—Analysis of Family Sector Energy and Future Trends* (Tokyo: IEE, July 27, 1981; in Japanese), p. 6.

41. Toichi, Adachi, and Yamamoto, "International Comparison," p. 12.

42. From 1973 to 1979, these variables increased at the following average annual percentages: population, 1.1; number of households, 2.2; number of occupied dwellings, 1.0; floor space of occupied dwellings, 0.8. See Statistics Bureau, *Japan Statistical Yearbook, 1980*, pp. 13, 454–55. Some numbers used in deriving these percentages were calculated or interpolated.

portable heaters to achieve higher room temperatures than are now customary. The fundamental question is whether traditional Japanese practices will persist, or whether higher incomes will lead to the adoption of more energy-intensive Western standards of comfort and heating systems. If such a shift does take place, its effect on energy consumption may be moderated through greater use of insulation, especially in new homes, and through the replacement of detached dwellings by multifamily units that have less exposed outer wall space.

The decline in the use of energy to cool space may also have been a delayed reaction to higher prices. In any event, some increase in the use of energy for this purpose is also possible. Much of Japan has a warm, humid climate for a good part of the year.[43] In 1979–80 about half of all Japanese households had air conditioners, but electricity consumption per unit was only about a fifth that in the United States.[44] Both Japanese air conditioners and the rooms that they cool are relatively small. It is also likely that Japanese households use their air conditioners more sparingly than do U.S. households. As in the case of space heating, higher incomes could lead to increased standards of comfort in warm weather and to the use of more energy for space cooling. More households may acquire air conditioners; some households may buy more than one unit; and both the size of units and the hours of use may increase.

Whether the decrease in energy used for cooking will continue at the rate recorded from 1973 to 1979 is difficult to predict. Some decrease is plausible, however, as increasingly prosperous families eat more meals away from home or more frequently use processed foods that require less cooking.

All of the above judgments concerning possible trends in the various components of residential energy consumption are entirely qualitative. The effort to express these judgments quantitatively in assumed future rates of change (last column of table 6-8) is therefore necessarily arbitrary. These assumptions replace the recent declines in the space heating and cooling categories with small rates of increase. The high recent rates of increase in energy used for water heating and lighting and electric power have been cut by more than half. The decline in the use of energy for cooking has been slightly increased. The total rate of change is a weighted average based on the 1979 functional breakdown.

43. In 1979–80 the national average number of degree days for Japan was 2,130, compared with 2,500 for the United States (both are based on 18 degrees Celsius). See Toichi, Adachi, and Yamamoto, "International Comparison," p. 4.

44. Ibid., p. 19.

The total assumed annual rate of future increase in table 6-8—3.1 percent—is not far from the 2.9 percent rate projected in case III (table 6-6) that assumed 1.0 percent annual declines in the ratios both of disposable income to GDP and of energy consumption to disposable income. An annual rate of increase of about 3.0 percent in residential energy consumption appears to be a reasonable choice for the central case during the 1980s. Acquisition of home appliances and water heating equipment by virtually all households in the 1980s would produce a lower rate of increase—say, 2.5 percent—in the 1990s. These rates produce the following projections (in millions of t.o.e.): 1990, 39.2; 1995, 44.4; 2000, 50.2.

COMMERCIAL AND MISCELLANEOUS ENERGY CONSUMPTION. Energy consumption in this part of the residential and commercial sector increased from 23.8 million t.o.e. in 1973, to 25.5 million t.o.e. in 1976, and to 27.8 million t.o.e. in 1979.[45] The annual rate of increase of 2.6 percent from 1973 to 1979 was substantially below the 3.3 percent annual rate of increase recorded by the residential part of the sector over the same period. Commercial and miscellaneous energy consumption increased somewhat more rapidly from 1976 to 1979 (2.9 percent a year) than it did from 1973 to 1976 (2.4 percent a year). The difference reflects the depressing effect of the 1973–74 oil crisis on economic activity in 1974 and 1975 (see table 2-1).

Table 6-9 gives figures for commercial and miscellaneous energy consumption in 1976 and 1979 and shows commercial consumption by principal users. Miscellaneous consumption was about the same fraction of the total in both years (23.1 percent in 1976 and 22.3 percent in 1979). Because the composition of miscellaneous consumption is so varied,[46] speculation about its future level is especially difficult. It will therefore be arbitrarily assumed that miscellaneous energy consumption will be about 20 percent of combined commercial and miscellaneous consumption in 1990.[47]

Commercial energy consumption (not including miscellaneous energy consumption) must be strongly influenced by the level of economic

45. These figures are the difference between total sectoral consumption shown in the energy balance tables and residential consumption.

46. Among other uses, miscellaneous energy consumption includes community, social, and personal services; electricity, gas, and water utilities (but not transformation losses); and government services.

47. This assumption takes into account that in recent years miscellaneous energy consumption has been increasing at a slower rate than commercial energy consumption.

Table 6-9. *Commercial and Miscellaneous Energy Consumption in Japan, 1976 and 1979*

	1976		1979		
Item	Ten trillions of kilo-calories[a]	Percentage	Ten trillions of kilo-calories[a]	Percentage	Annual rate of change (percent)
Offices and buildings	4.7	18.4	5.4	19.4	4.7
Wholesale and retail business	2.7	10.6	3.0	10.8	3.6
Hospitals	2.6	10.2	2.8	10.1	2.5
Hotels	2.5	9.8	2.8	10.1	3.9
Schools	2.3	9.0	2.4	9.0	2.8
Restaurants	1.7	6.7	2.0	7.2	5.6
Laundries, barbershops, and bathhouses	1.6	6.3	1.7	6.1	2.0
Department stores and supermarkets	0.8	3.1	0.9	3.2	4.0
Theaters, movie houses, and other entertainment	0.6	2.4	0.6	2.2	0.0
Subtotal (total commercial)	19.6	76.9	21.6	77.7	3.3
Miscellaneous[b]	5.9	23.1	6.2	22.3	1.7
Total	25.5	100.0	27.8	100.0	2.9

Source: Energy Conservation Research Center, "National Living Standards," pp. 4–5, 12. Numbers are rounded.
a. Ten trillion kilocalories approximately equal a million t.o.e.
b. Miscellaneous includes, among other uses, community, social, and personal services; electricity, gas, and water utilities (but not transformation losses); and government services. Miscellaneous consumption is the difference between the sectoral total given in the energy balance tables and the sum of residential and commercial consumption.

activity in the commercial part of the economy, which is measured by value added in commercial activities. Given the assumptions that GDP will grow at an average annual rate of 5.0 percent and that energy prices will remain roughly constant in real terms, future commercial energy consumption will depend upon the share of commerce in GDP (the ratio of value added to GDP) and the intensity of energy use in commercial activities (the ratio of energy consumption to value added).

The share of commerce in GDP varies slightly from year to year, but no clear trend is apparent. It was 55.3 percent of GDP in 1973, 56.2 percent in 1976, and 55.0 percent in 1979. Assuming that it will be about 55.0 percent of GDP in 1990 is as defensible as any other assumption. The intensity of commercial energy use, measured by the ratio of commercial energy consumption to value added in commerce, was 2.24 kilocalories per yen in 1976 and 2.16 kilocalories per yen in 1979—an annual rate of decline of about 1.2 percent.[48]

48. Estimates of commercial energy consumption in 1976 and 1979 were drawn from Energy Conservation Research Center, "National Living Standards," pp. 4–5, 12. GDP and value added in commerce (both in constant 1975 yen) are from OECD, *National Accounts of OECD Countries, 1962–79*, vol. 2 (Paris: OECD, 1981), p. 33.

If this rate of decline in energy intensity (measured by the ratio of energy consumption to value added) continues—under the stated assumptions concerning GDP and the ratio of value added to GDP—commercial energy consumption would be 32.7 million t.o.e. in 1990. If the intensity of energy use remained constant at its 1979 level, however, commercial energy consumption would be 37.3 million t.o.e. in 1990. An increase in energy intensity is theoretically possible but appears unlikely in light of recent trends. Constant intensity (the second set of projections set forth above) can probably be taken as a limiting upper case. The real question appears to be how rapidly energy intensity will decline. Assuming a continuation of the 1976–79 rate of decline in intensity (the first set of projections set forth above) is not an unreasonable choice for the central or reference case.

The annual rate of increase in commercial energy consumption from 1979 to 1990 under these projections is 3.8 percent. In its scenario for low oil prices the IEE arrived at almost the same result: 3.7 percent.[49] The IEE projected only a 2.1 percent annual rate of increase in commercial energy consumption from 1990 to 2000. This reduction in the rate of increase in commercial energy consumption is associated, however, with a slowing down in the projected annual GDP growth rate from 4.7 percent in the period 1979–90 to 3.7 percent in the period 1990–2000. If GDP is assumed to increase 5.0 percent annually, a somewhat higher rate of increase in commercial energy consumption would be justified.

Even in the absence of further increases in real energy prices, the decline in the intensity of commercial energy use may well continue through the 1990s. Upgrading the energy efficiency of the capital stock of commercial enterprises will be a slow process. The following breakdown of commercial energy use in 1976 gives some idea of what is involved:[50]

	Percentage
Heating space	29.9
Cooling space	6.8
Heating water	27.9
Cooking	5.8
Lighting, motive power, and other electric power	29.6

Accelerated replacement of free-standing and relatively short-lived

49. Energy Conservation Research Center, "National Living Standards," p. 12.
50. Ibid., p. 5.

Table 6-10. *Projected Residential, Commercial, and Miscellaneous Energy Consumption in Japan, Selected Years, 1979–2000, Central Case*
Ten trillions of kilocalories[a]

Year	Residential	Commercial	Miscellaneous	Total
1979 (actual)	28.3	21.6	6.2	56.1
1990	39.2	32.7	8.2	80.1
1995	44.4	39.3	9.8	93.5
2000	50.2	48.2	12.0	110.4

Sources: See the text. Numbers are rounded.
a. Ten trillion kilocalories approximately equal a million t.o.e.

kitchen equipment and electrical appliances might take place to cut energy costs. But structural alterations (for example, new lighting systems) and replacement of large, expensive, and long-lasting equipment (for example, new boilers to heat space and water) might not be accelerated. Over a longer period of time, however, normal schedules of equipment replacement and the construction of new buildings would continuously improve energy efficiency.

The figures shown above indicate one area of vulnerability: the relatively small portion of commercial energy consumption devoted to the cooling of space. The possibility of a wider use of air conditioning in Japanese commercial establishments justifies assuming a somewhat lower decline in energy intensity during the 1990s than was assumed for the 1980s.

If the intensity of commercial energy use declines 1.0 percent annually during the 1990s (and if previously stated assumptions concerning GDP and the ratio of value added to GDP are applied), commercial energy consumption would be 39.3 million t.o.e. in 1995 and 48.2 million t.o.e. in 2000. The annual rate of growth in such consumption from 1990 to 2000 would be 2.6 percent.

SECTORAL ENERGY CONSUMPTION. The components of the central case projections of energy consumption in the residential and commercial sector can be summed up as in table 6-10. Miscellaneous energy consumption in future years has been arbitrarily set at 25 percent of commercial consumption. The projections for 1990 in table 6-10 are quite close to those calculated by the IEE in May 1981.[51]

51. See note 3, above, and source note to table 6-1. The IEE projected 77.5 million t.o.e. for residential and commercial energy consumption in 1990. The assumptions underlying the IEE projection are given in the notes to table 6-1.

In the period 1976–79, a 2.35 percent increase in energy consumption in the residential and commercial sector was associated with a 6.01 percent increase in real GDP. The average income elasticity of residential and commercial energy demand during that period was therefore 0.39. The above projections have residential and commercial energy consumption increasing at an average annual rate of 3.38 percent during the period 1979–2000. Given the assumed 5.0 percent annual increase in GDP, the implied income elasticity is 0.68.

The ratio between residential and commercial energy consumption and GDP is also projected to change. In 1979 this ratio was 3.0 kilocalories per 1975 yen. Under the projections listed above and under the assumed 5.0 percent annual growth in real GDP, the ratio declines as follows: 1990, 2.5; 1995, 2.3; 2000, 2.2. In the year 2000 the amount of residential and commercial energy consumption associated with a unit of GDP would be only about three-fourths of what it was in 1979.

Transportation Sector

In estimating future energy consumption in the transportation sector, freight transportation and passenger transportation will be treated separately. Data for calendar years will be used because the energy balance tables for fiscal years do not list energy consumed in freight and passenger transportation separately.

ENERGY CONSUMED IN FREIGHT TRANSPORTATION. Energy requirements for freight transportation depend in large part on the general level of economic activity. In principle it should therefore be possible to project future energy consumption in freight transportation by estimating the ratio of such consumption to GDP and by assuming a rate of growth for GDP. The ratio of energy consumed in freight transportation to GDP is not fixed, however. It varies in response to changes in the structure of the economy, the mix of modes of transportation, and the intensity of energy use in each of those modes.

In thinking about future energy consumption in freight transportation, it is useful to focus on two variables: the number of ton-kilometers of freight to be moved per unit of GDP and the amount of energy required per ton-kilometer. The first of these variables is, of course, structural; the second measures the intensity of energy use. Table 6-11 shows the changes in these variables during the period 1965–79. Values for the various modes of transportation are also shown. A clearer view of past

Table 6-11. *Energy Consumed in Freight Transportation in Japan, Selected Years, 1965–79*

Year	Thousands of kilocalories per metric ton-kilometer					Ton-kilometers per thousand 1975 yen of GDP			
	Trucks	Railways	Ships	Aviation	Total	Trucks	Railways	Ships and aviation	Total
1965	1.573	0.417	0.250	10.000	0.646	0.699	0.828	1.166	2.693
1970	1.064	0.311	0.254	9.919	0.580	1.153	0.537	1.284	2.976
1971	1.038	0.288	0.299	8.642	0.590	1.157	0.505	1.274	2.937
1972	1.073	0.239	0.250	8.621	0.576	1.145	0.444	1.312	2.901
1973	1.303	0.201	0.251	8.333	0.611	0.966	0.400	1.424	2.790
1974	1.325	0.202	0.319	8.571	0.656	0.905	0.363	1.333	2.602
1975	1.375	0.186	0.347	8.618	0.699	0.877	0.320	1.243	2.441
1976	1.449	0.181	0.322	8.250	0.708	0.852	0.297	1.249	2.399
1977	1.434	0.198	0.312	7.650	0.718	0.873	0.252	1.236	2.362
1978	1.366	0.206	0.303	7.589	0.702	0.907	0.239	1.233	2.380
1979	1.290	0.195	0.283	6.996	0.672	0.952	0.237	1.245	2.434
					Average annual growth rates (percent)				
1965–73	−2.33	−8.72	0.05	−2.25	−0.69	4.13	−8.69	2.53	0.44
1973–79	−0.17	−0.50	2.02	−2.87	1.60	−0.24	−8.35	−2.21	−2.25
1976–79	−3.80	2.51	−4.21	−5.35	−1.72	3.77	−7.25	−0.11	0.48

Sources: Energy figures are from IEE, *Energy Balance Tables* (Tokyo: IEE, 1982 = ton-kilometer data are from Bureau of Statistics, *Japan Statistical Yearbook, 1973/74*, p. 260; *1977*, p. 258; *1980*, p. 258; *1981*, p. 271; GDP data are from OECD, *National Accounts, 1962–79*, vol. 2, p. 33. Numbers are rounded.

Table 6-12. *Indexes of Energy Consumed in Japanese Freight Transportation, 1970–79*
1965 = 100

Year	Energy per ton-kilometer	ton-kilometer per yen of real GDP
1970	89.8	110.5
1971	91.3	109.1
1972	89.2	107.7
1973	94.6	103.6
1974	101.5	96.6
1975	108.2	90.6
1976	109.6	89.1
1977	111.1	87.7
1978	108.7	88.4
1979	104.0	90.4

Source: Table 6-11. Numbers are rounded.

developments can be obtained by converting the absolute figures in table 6-11 to index numbers, as shown in table 6-12. The figures in table 6-12 show no well-defined trends for either variable for the entire period 1965–79.

From 1965 to 1972 freight transportation became less energy intensive. From 1973 to 1977, it became more energy intensive, and in 1978 and 1979 energy intensity declined again. The amount of freight transportation required per yen of real GDP followed roughly an opposite path—rising from 1965 to 1971, falling from 1972 to 1977, and rising in 1978 and 1979. These inverse movements may be related. Rising freight requirements mean fuller loads in trucks, trains, and ships and greater efficiency of energy use, and declining energy requirements may mean less full loads and lower energy efficiency.

In projecting possible future energy consumption, recent trends in energy intensity and freight requirements appear to be more useful than changes in these variables over a longer period, such as 1973–79. From 1976 to 1979 annual changes were in one direction with no reversals; the rate of growth of real GDP was also fairly steady (see table 2-1).

As might be expected, economic conditions from 1976 to 1979 did not conform exactly to the assumptions of the central case. GDP grew at an average annual rate of 5.7 percent, rather than the 5.0 percent assumed. Rather than remaining stable, real prices of transportation fuels may have declined.[52] The practical significance of these differences between actual conditions and the assumptions of the central case is probably not large. It is possible that greater requirements for freight transportation per yen of GDP are generated at high than at low rates of GDP growth,[53] but a differential of 0.7 percent could not have much effect on the average relationship between ton-kilometers and GDP in the next few years. In the short run, lower energy prices would not have much effect on requirements for freight transportation per yen of GDP.[54] Lower energy prices would weaken incentives to reduce the intensity of energy use, but the short-run effects on energy intensity would probably be small because the stock of transport vehicles does not turn over rapidly.

The differences between actual conditions and the assumptions of the central case do not appear to rule out using 1976–79 trends in projecting future energy consumption in freight transportation. A more serious problem is whether it is reasonable to expect that those trends will continue many years into the future.

This problem is complicated by the different behaviors of the various modes of transportation in the period 1976–79. Highway, sea, and air transportation became less energy intensive, but railways became more so. The ratio between highway transportation requirements and GDP increased, but the ratio between railway transportation requirements and GDP fell. (That is, the shift from trains to trucks continued.) The ratio of sea and air transportation requirements and GDP changed very little. Trends for the 1976–79 period provide as good a basis as any other for the central case projection of energy consumption in freight transportation in 1985. After 1985 the possibility that some of these trends may change must be given serious consideration.

Table 6-13 presents three illustrative projections of energy consumption in freight transportation in 1985 and 1990. All three cases assume a 5.0 percent annual growth of real GDP and little or no change in real energy prices. The 1985 projections use 1976–79 trends in the ratios of ton-kilometers to GDP and of energy consumption to ton-kilometers.

52. The real price of regular gasoline fell about 13.5 percent from 1976 to 1978. Information on other energy prices paid by providers of freight transportation is not available.

53. This would be the case if sectors that generate a disproportionate share of requirements for freight transportation grew faster relative to the economy as a whole at high GDP growth rates than at low GDP growth rates.

54. Over the long run, energy prices would affect requirements for freight transportation by influencing the location of various economic activities.

Table 6-13. *Energy Consumed in Freight Transportation in Japan,*
1979 and Illustrative Projections for 1985 and 1990

Year and mode	Millions of ton-kilometers	Ton-kilometers per thousand 1975 yen of GDP[a]	Kilocalories per ton-kilo-meter[b]	Energy con-sumption (ten billions of kilo-calories)[c]
1979				
Truck	172,888	0.952	1,290	22,309
Rail	43,087	0.237	195	840
Sea, air	226,063	1.245	291	6,577
Total	442,038	2.434	672	29,726
1985[d]				
Truck	289,214	1.189	1,023	29,579
Rail	36,729	0.151	226	830
Sea, air	300,889	1.237	229	6,890
Total	626,832	2.577	595	37,299

1990 (Case I: assumes 1976–79 trends in ton-kilometers/GDP and energy/ton-kilometer continue in 1985–90)[e]

Truck	443,935	1.430	843	37,408
Rail	32,286	0.104	256	826
Sea, air	381,846	1.230	188	7,163
Total	858,067	2.764	529	45,397

1990 (Case II: assumes 1976–79 trends in energy/ton-kilometer continue in 1985–90; no change in ton-kilometers/GDP)[e]

Truck	369,118	1.189	843	31,117
Rail	46,877	0.151	256	1,200
Sea, air	384,019	1.237	188	7,220
Total	800,014	2.577	494	39,537

1990 (Case III: assumes half 1976–79 rates of change in energy/ton-kilometer and ton-kilometers/GDP in 1985–90)[e]

Truck	405,129	1.305	929	37,636
Rail	39,116	0.126	241	943
Sea, air	382,777	1.233	207	7,923
Total	827,022	2.664	562	46,502

Sources: Ton-kilometer data are from Statistics Bureau, *Japan Statistical Yearbook, 1981*, p. 271; GDP data are from OECD, *National Accounts, 1962–79*, vol. 2, p. 33; energy data are from IEE, *Energy Balance Tables* (1982). Numbers are rounded.

a. GDP was 181,510 billion yen in 1979. Under the assumed 5.0 percent annual growth, it would be 243,241 billion yen in 1985 and 310,444 billion yen in 1990. All yen figures are constant 1975 yen.

b. The annual rates of growth of ton-kilometers per GDP from 1976 to 1979 were: truck, 3.77 percent; rail, −7.25 percent; sea and air, −0.11 percent. The annual rates of growth of energy (kilocalories) per ton-kilometer from 1976 to 1979 were: truck, sea and air, −0.11 percent. The annual rates of growth of energy per ton-kilometer from 1976 to 1979 were: truck, −3.80; rail, 2.51; sea and air, −3.91.

c. Ten billion kilocalories approximately equal a thousand t.o.e.

d. Projections for 1985 assume that 1976–79 trends in the ratios of ton-kilometers to GDP and of energy consumption to ton-kilometers will continue, that real GDP growth will be 5.0 percent, and that energy prices will remain stable.

e. All cases assume 5.0 percent annual increase in real GDP and little or no change in real energy prices.

Case I uses these 1976–79 trends in projecting energy consumption in 1990. Case II carries the 1976–79 trends in energy intensity forward to 1990 but assumes that the ratio of ton-kilometers to GDP will be the same in 1990 as in 1985 (that is, that the shift from trains to trucks will end in 1985). Case III assumes that the forces that drove the 1976–79 trends will eventually weaken and therefore uses rates of change one-half those of 1976–79 in projecting energy consumption in 1990.

Case II indicates the importance of the shift from trains to trucks in retarding the decline in energy intensity. Cases I and III illustrate the opposing effects of falling energy intensity and rising freight transportation requirements per yen of GDP. Although the rates of change in these variables are half as great in case III as in case I, the resultant energy consumption is quite similar.

Because a tapering off of 1976–79 trends is a plausible hypothesis, case III will be used as the central case projection of energy consumption in freight transportation. Energy consumption in freight transportation is therefore projected to increase at an annual rate of 4.5 percent from 1985 to 1990, compared with the annual rate of increase of 3.9 percent projected for the period 1979–85.

For several reasons, an annual rate of increase as high as 4.5 percent in the 1990s appears unlikely under the assumptions of economic growth and energy prices used in the central case. The shift away from energy-efficient railways may well slow down and could even be reversed. The increased emphasis on the use of advanced technologies could increase the average value of a ton of product, thereby reducing the ton-kilometers to be transported per unit of GDP. Moreover, past increases in energy prices could continue to provide incentives to replace existing freight transportation systems with more energy-efficient ones and to give greater weight to energy costs in locating economic activities.

Estimating the quantitative effects of the various influences that may slow down the rate of increase of energy consumption in freight transportation is impossible. Because of the strong reasons to expect that such a decline will occur, the rate of increase will arbitrarily be set at 3.5 percent a year from 1990 to 1995 and 2.5 percent a year from 1995 to 2000. Energy consumption in freight transportation would then be 55.2 million t.o.e. in 1995 and 62.4 million t.o.e. in 2000.

ENERGY CONSUMED IN PASSENGER TRANSPORTATION. Requirements for passenger transportation reflect both the general level of economic activity and decisions by individuals concerning the expenditure of disposable income.[55] The relative importance of these two influences can only be a subject for speculation. Fortunately, for purposes of the present analysis the relationship between disposable income and GDP is fairly stable, so regarding GDP as the fundamental determinant of passenger transportation requirements appears to be an acceptable simplification.[56] Given requirements for passenger transportation, energy consumption depends on the intensity of energy use in the various modes of transportation and on the relative importance of those modes.

Table 6-14 shows how the intensity of energy use and the relationship between passenger transportation requirements and GDP have varied over time with respect to the modes of transportation. Although energy intensity in each of the modes declined, total intensity rose steadily. This seeming paradox is explained by the growing importance of cars, a relatively energy-intensive means of transportation. The decline in passenger-kilometers per unit of real GDP is not surprising. Japan's GDP has been increasing much more rapidly than the size of its labor force.[57] Adding workers increases passenger transportation requirements, but raising the productivity of the existing labor force has no direct effect on such requirements. The influence of rising disposable incomes has not been strong enough to enable transportation requirements to keep pace with GDP.

Table 6-15 presents three illustrative projections of energy consumption in passenger transportation in 1985 and 1990. These projections parallel those for energy consumption in freight transportation presented in table 6-13. The 1985 projections use 1976–79 trends in energy intensity (the ratio of energy consumption to passenger-kilometers) and in the relationship between passenger-kilometers traveled and real GDP. Case I projects the same trends into 1990. Case II uses 1976–79 trends in energy intensity for 1990 estimates but assumes no change after 1985 in the ratio of passenger-kilometers to GDP. Case III assumes that both

55. The demand for transportation services also presumably varies inversely with the real cost of such services. Estimating both possible future cost changes and the price elasticity of demand for each mode of transportation would, however, be quite difficult. In any case, the energy component of the cost of transportation has been assumed to remain constant.

56. This approach may introduce a small upward bias in projections of requirements for passenger transportation, since disposable income has shown a tendency to decline as a percentage of GDP.

57. From 1965 to 1979 the labor force grew at an average annual rate of 1.03 percent, whereas GDP increased at an average annual rate of 7.8 percent.

Table 6-14. *Energy Consumed in Passenger Transportation in Japan, 1965–79*

	Kilocalories per passenger-kilometer					Passenger-kilometers per thousand 1975 yen of GDP				
Year	Car	Bus	Rail	Other	Total	Car	Bus	Rail	Other	Total
1965	1,041.87	120.85	64.80	998.33	195.05	0.587	1.158	3.691	0.087	5.522
1970	590.24	129.35	59.42	709.86	251.28	1.538	0.873	2.451	0.120	4.983
1971	608.55	131.94	57.86	704.58	274.63	1.717	0.818	2.353	0.124	5.011
1972	618.47	139.00	54.61	913.43	277.58	1.642	0.807	2.239	0.145	4.832
1973	714.36	150.67	54.84	584.62	310.06	1.546	0.765	2.144	0.160	4.616
1974	710.51	138.51	52.79	588.84	303.14	1.581	0.802	2.244	0.174	4.801
1975	707.22	147.32	48.64	641.15	318.00	1.697	0.745	2.191	0.176	4.808
1976	704.88	165.55	51.21	634.96	332.76	1.699	0.634	2.053	0.171	4.557
1977	730.45	156.02	54.08	613.95	343.92	1.611	0.638	1.906	0.184	4.340
1978	709.53	149.53	56.93	611.41	353.33	1.720	0.622	1.808	0.194	4.343
1979	699.50	149.58	57.25	615.30	360.71	1.762	0.597	1.722	0.202	4.282
					Average annual growth rates (percent)					
1965–73	4.61	2.80	-2.06	-6.47	5.97	12.87	-5.05	-6.56	7.19	-2.22
1973–79	-0.35	-0.12	0.72	0.86	2.55	2.20	-4.05	-3.59	3.96	-1.24
1976–79	-0.26	-3.32	3.79	-1.04	2.72	1.22	-1.98	-5.69	5.71	-2.05

Sources: Energy data are from IEE, *Energy Balance Tables (1982)*; passenger-kilometer data are from Statistics Bureau, *Japan Statistical Yearbook, 1973/74*, p. 261; *1980*, p. 259; *1981*, p. 271; GDP data are from OECD, *National Accounts, 1962–79*, vol. 2, p. 33. Numbers are rounded.

Table 6-15. *Energy Consumed in Passenger Transportation in Japan, 1979 and Illustrative Projections for 1985 and 1990*

Year and mode	Hundred millions of passenger-kilometers	Passenger-kilometers per thousand 1975 yen of GDP[a]	Kilocalories per passenger-kilometer[b]	Energy consumption (ten billions of kilocalories)[c]
1979				
Car	3,199	1.762	700	22,377
Bus	1,083	0.597	150	1,620
Rail	3,125	1.722	57	1,789
Other	366	0.201	616	2,252
Total	7,773	4.282	361	28,038
1985[d]				
Car	4,609	1.895	689	31,742
Bus	1,287	0.529	122	1,572
Rail	2,948	1.212	72	2,109
Other	681	0.280	578	3,940
Total	9,525	3.916	413	39,363

1990 (Case I: assumes 1976–79 trends in passenger-kilometers/GDP and energy/passenger-kilometer continue in 1985–90)[e]

Year and mode	Hundred millions of passenger-kilometers	Passenger-kilometers per thousand 1975 yen of GDP[a]	Kilocalories per passenger-kilometer[b]	Energy consumption (ten billions of kilocalories)[c]
Car	6,249	2.013	680	42,479
Bus	1,487	0.479	103	1,534
Rail	2,806	0.904	86	2,417
Other	1,149	0.370	549	6,309
Total	11,691	3.766	451	52,739

1990 (Case II: assumes 1976–79 trends in energy/passenger-kilometer continue in 1985–90, no change in passenger-kilometers/GDP)[e]

Year and mode	Hundred millions of passenger-kilometers	Passenger-kilometers per thousand 1975 yen of GDP[a]	Kilocalories per passenger-kilometer[b]	Energy consumption (ten billions of kilocalories)[c]
Car	5,883	1.895	680	40,004
Bus	1,642	0.529	103	1,691
Rail	3,762	1.212	86	3,235
Other	869	0.280	549	4,771
Total	12,156	3.916	409	49,701

1990 (Case III: assumes half 1976–79 rates of change in energy/passenger-kilometer and passenger-kilometers/GDP in 1985–90)[e]

Year and mode	Hundred millions of passenger-kilometers	Passenger-kilometers per thousand 1975 yen of GDP[a]	Kilocalories per passenger-kilometer[b]	Energy consumption (ten billions of kilocalories)[c]
Car	6,066	1.954	684	41,491
Bus	1,562	0.503	112	1,749
Rail	3,256	1.049	79	2,572
Other	1,000	0.322	563	5,630
Total	11,884	3.828	433	51,442

Sources: Passenger-kilometer data are from Statistics Bureau, *Japan Statistical Yearbook, 1981*, p. 271; GDP data are from OECD, *National Accounts, 1962–79*, vol. 2, p. 33; energy data are from IEE, *Energy Balance Tables (1982)*. Numbers are rounded.

a. For GDP figures, see notes to table 6-13.

b. The annual rates of growth of passenger-kilometers per GDP from 1976 to 1979 were: car, 1.22 percent; bus, −1.98 percent; rail, −5.69 percent; other, 5.71 percent. The annual rates of growth of energy (kilocalories) per passenger-kilometer from 1976 to 1979 were: car, −0.26 percent; bus, −3.32 percent; rail, 3.79 percent; other, −1.04 percent.

c. Ten billion kilocalories approximately equal a thousand t.o.e.

d. Projections for 1985 assume that 1976–79 trends in the ratios of passenger-kilometers to GDP and of energy consumption to passenger-kilometers will continue, that GDP growth will be 5.0 percent, and that energy prices will remain stable.

e. All cases assume 5.0 percent annual increase in real GDP and little or no change in real energy prices.

Table 6-16. *Projected Energy Consumption in Japanese Freight and Passenger Transportation, Selected Years, 1979–2000, Central Case*
Ten billions of kilocalories[a]

Year	Freight	Passenger	Total
1979[b]	29.7	28.0	57.7
1985	37.3	39.4	76.7
1990	46.5	51.4	97.9
1995	55.2	64.0	119.2
2000	62.4	76.0	138.4

Source: See the text. Numbers are rounded.
a. Ten billion kilocalories approximately equal a thousand t.o.e.
b. Actual figures.

energy intensity and the ratio of passenger-kilometers to GDP will change from 1985 to 1990 at half the rates recorded in 1976–79.

To provide symmetry with the treatment of freight transportation, and because a tapering off of 1976–79 trends is again plausible, case III will be used as the central case projection of energy consumed in passenger transportation. Under this projection, energy consumption grows at an annual rate of 5.5 percent from 1985 to 1990. Projecting this rate of growth into the 1990s, however, would probably not be realistic. As in freight transportation, the shift away from energy-efficient railways may come to an end. The increase in the number of cars may also taper off as crowded streets and highways make car ownership less attractive. It is even possible that over the long run high transportation costs will induce workers to move closer to their places of of employment. Because of these influences working toward a slower rate of growth in energy consumed in passenger transportation, the annual rate of increase will be arbitrarily set at 4.0 percent for the period 1990–95 and at 3.0 percent for the period 1995–2000. Energy consumed in passenger transportation would then be 64.0 million t.o.e. in 1995 and 76.0 million t.o.e. in 2000.

SECTORAL ENERGY CONSUMPTION. The central case projections of energy consumption in transportation can be summed up as in table 6-16. The income elasticities of energy demand implicit in the sectoral projections for 1985 and 1990 are not markedly different from those of recent years. The projections for 1995 and 2000, however, imply sharp reductions in income elasticity. The average income elasticities for the periods indicated below are: 1973–79, 1.059; 1979–85, 0.960; 1985–90, 1.000; 1990–95, 0.804; 1995–2000, 0.606.

Under these projections, the ratio of energy consumed in transportation to GDP (expressed in kilocalories per 1975 yen) also would not change much until the 1990s: 1973, 3.14; 1979, 3.18; 1985, 3.15; 1995, 3.01; 2000, 2.74.

The central case projections call for a 4.8 percent average annual rate of increase in final energy consumption from 1979 to 1990. This rate is higher than those in two other recent projections (there appear to be no comparable projections for the period 1990–2000). Estimates of the IEE made in May 1981 (cited previously in this chapter), show an annual rate of increase of only 1.8 percent. Official Japanese government figures published by the OECD (adjusted to make them comparable) show an annual rate of increase of 3.8 percent.[58]

The central case projections are consistent with the experience of 1976 to 1979, when energy consumed in transportation increased at an annual rate of 4.9 percent. The adjusted official estimates are close to the 3.9 percent rate of increase experienced from 1973 to 1979. The IEE projection is lower than the others principally because it used more optimistic assumptions concerning the energy efficiency of highway transportation.[59]

Total Future Final Energy Consumption

The only component of total future energy consumption by final consumers not yet considered is nonenergy uses of fuels. This category is very small in Japanese energy statistics and comprises uses of lubricants, grease, asphalt, and certain other petroleum products. Chemical feedstocks are included in the industrial sector rather than in the nonenergy sector. From 1974 to 1979, nonenergy uses of energy materials constituted a fairly constant percentage of total final energy consumption, with the average for the period being 2.5 percent. It will be assumed that nonenergy consumption will continue to be 2.5 percent of total final energy consumption in the future.

Total final energy consumption in the central case can be summed up as in the top part of table 6-17. The combined effect of the sectoral projections of energy consumption causes total final energy consumption to increase at an annual rate of 3.0 percent from 1979 to 1990 and at an

58. OECD, International Energy Agency (IEA), *Energy Policies and Programmes of IEA Countries, 1980 Review* (Paris: OECD, 1981), pp. 190–91. The adjustments made are described in the source note to table 6-18.
59. Conversation at the IEE (Tokyo, February 1982).

Table 6-17. *Disposition and Structure of Future Final Energy Consumption in Japan, Selected Years, 1979–2000, Central Case*

Sector	1979[a]	1990	1995	2000
	Disposition (ten trillions of kilocalories)[b]			
Industrial	150.1	188.8	206.4	225.9
Residential and commercial	56.1	80.1	94.5	112.8
Transportation[c]	57.9	97.9	119.2	138.4
Nonenergy uses of fuels	7.1	9.4	10.8	12.2
Total	271.2	376.2	430.9	489.3
	Structure (percent)			
Industrial	55.3	50.2	47.9	46.2
Residential and commercial	20.7	21.3	21.9	23.0
Transportation	21.3	26.0	27.7	28.3
Nonenergy uses of fuels	2.6	2.5	2.5	2.5
Total	100.0	100.0	100.0	100.0

Sources: See the text. Numbers are rounded.
a. Fiscal year (beginning April 1).
b. Ten trillion kilocalories approximately equal a million t.o.e.
c. Energy consumed in 1979 was 57.7 million t.o.e., an insignificant difference from the fiscal year figure shown.

annual rate of 2.7 percent from 1990 to 2000. The sectoral projections are based on trends and relationships that emerged in the years between the two oil crises. The extent to which those trends and relationships may be changed by Japan's adjustment to the second oil crisis and other international developments is not yet clear.

These projections also show changes in the structure of energy consumption, as shown in the bottom part of table 6-17. The major structural changes, which are the results of different sectoral rates of growth, are the decline in the share of the industrial sector in total final energy consumption and the rise in the shares of the residential and commercial sector and the transportation sector. (The constant share of the nonenergy sector has no special significance because this constancy was assumed.)

Alternatives to the Central Case

Energy consumption varies directly with the rate of growth of real GDP and inversely with the rate of change in real energy prices. Estimates of the responsiveness (that is, the elasticity) of energy consumption to changes in GDP are implicit in the central case projections. These estimates can be used in constructing alternatives to the central case.

Table 6-18. *Estimated Energy Consumption in Japan in 1990 under Different Assumptions Concerning the Annual Rate of Growth of GDP*
Ten trillions of kilocalories[a]

	GDP growth rate (percent)		
Sector	4	5	6
Industrial	180.5	188.8	197.5
Residential and commercial	74.5	80.1	85.8
Transportation	88.3	97.9	108.4
Nonenergy uses of fuels	8.9	9.4	9.9
Total	352.2	376.2	401.6

Sources: The estimates of energy consumption under the assumption of 5.0 percent annual GDP growth are the central case projections described in the text. The estimates of energy consumption under the assumption that GDP will increase annually at 4.0 percent or 6.0 percent were calculated by using the GDP elasticities implicit in the central case projections.

a. Ten trillion kilocalories approximately equal a million t.o.e.

Because the central case assumed relatively stable real energy prices, however, no price elasticities of energy demand are implicit in the central case projections. An idea of the price elasticity of energy demand can be obtained only by analyzing past experience.

Table 6-18 compares the central case projections for 1990 with projections that assume higher or lower annual rates of growth of GDP. For illustrative purposes, GDP growth rates of 4.0 percent and 6.0 percent were chosen. Sectoral energy consumption under these assumptions was calculated by using the GDP elasticities implicit in the central case projections.[60]

Because sectoral GDP elasticities differ, the structure of energy consumption is affected by the rate of growth of GDP. Since the GDP elasticity of energy consumption by industry is relatively low, the share of the industrial sector in total final energy consumption declines as the GDP growth rate rises. The reverse is true of the transportation sector, which has a relatively high GDP elasticity of demand for energy.

The overall GDP elasticity implicit in the central case projection is about 0.6. This overall GDP elasticity is substantially below the elasticity that prevailed in the past. A regression analysis of changes in GDP,

60. This method could be used to derive alternatives to the central case projections for 1995 and 2000. Alternatives are not presented here because those projections would be even more speculative than the central case projections for 1990.

energy consumption, and fuel and energy prices over the period 1965–79 yielded a long-term GDP elasticity of about 1.0.[61]

The same regression analysis yielded a long-term price elasticity of about −0.5. Price elasticity may, of course, be higher or lower in the future, but the past elasticity can illustrate how changes in real energy prices could alter the total final energy consumption projected in the central case. (Comparable sectoral price elasticities are not available.)

Total final energy consumption in 1990 can be shown under three different assumptions of real energy prices. Case A assumes that energy prices rise 1.0 percent a year. Case B assumes that energy prices fall 1.0 percent a year. The central case, of course, assumes relatively stable energy prices. In all three cases GDP is assumed to increase at an average annual rate of 5.0 percent, and price elasticity is assumed to be −0.5. Total final energy consumption in 1990, in millions of t.o.e., would be: 355.2 (case A), 376.2 (central case), and 396.6 (case B).

Future rates of change in real GDP and real energy prices may diverge more from the assumptions of the central case than do any of the illustrative cases set forth above. Possible price developments are grounds for particular uncertainty. In 1980, in a delayed response to the doubling of the international price of crude oil in 1979, the fuel and energy price index (deflated by the wholesale price index) rose 39.8 percent above its 1979 level.[62]

The quarterly rate of increase in the deflated fuel and energy price index, compared with the same quarter of the previous year, peaked in the second quarter of 1980 and declined in the third and fourth quarters of 1980 and in the first quarter of 1981.[63] More recent data are not

61. Double-log, time-series regression analysis was done on data from 1965 to 1979 using a variation of the Koyck lag-adjustment model. See L. M. Koyck, *Distributed Lags and Investment Analysis* (Amsterdam: North-Holland, 1954), which specifies total final energy consumption of the Japanese economy as a function of GDP (in constant 1975 yen), price of energy measured by an index of fuel and energy prices to Japanese producers (deflated using the wholesale price index, 1975 = 100), and lagged energy consumption. The adjusted R_2 was 0.995, and the Durbin-Watson statistic was 2.190. Estimated coefficients were all significant at a 1 percent level or better. The results were a short-term GDP elasticity of 0.59 and long-term elasticity of 1.12. The short-term price elasticity was −0.28, the long-term −0.54. The energy data used in the regression analysis were from IEE, *Energy Balances, 1982* (Tokyo: IEE, 1982). GDP figures were from OECD, *National Accounts, 1962–79*, vol. 2, p. 33. The fuel and energy index and the wholesale price index were from Statistics Bureau, *Japan Statistical Yearbook, 1970*, p. 367; *1973/74*, p. 365; and *1981*, pp. 390–91.

62. *OECD Economic Surveys: Japan*, July 1981, p. 16.

63. Ibid.

available, but it is possible that the depressed state of the world economy and the softening of international oil prices in the early 1980s have caused real energy prices to decline, at least in the near term.

How much, if any, of the increase in real energy prices in 1980 and early 1981 may be wiped out by price declines is difficult to judge. Nor can the course of oil prices over the long run, and their effects on energy prices paid by final consumers, be foreseen. If the trend in real energy prices is upward, energy consumption will be depressed, and the projections of the central case will be too high—by how much will depend on the rate of increase in real energy prices and the price elasticity of energy demand.

CHAPTER SEVEN

Means of Meeting Japan's Future Energy Requirements

THE FUTURE energy requirements of final consumers in Japan were considered in the preceding chapter. This chapter examines the magnitude, composition, and geographical origin of the primary energy supplies that will be needed to meet those requirements. Some of the fuels that go into final energy consumption—in particular, electricity and refined petroleum products—are not primary energy supplies but secondary products whose creation consumes energy. Estimating future primary energy supplies therefore involves two questions. First, what fuels, both primary and secondary, are likely to be used to meet the requirements of each consuming sector? Second, what primary energy supplies must be used to produce the secondary fuels consumed? Once the magnitude and composition of the primary energy supplies that Japan will need have been estimated, two further questions can be addressed. How much primary energy can Japan produce domestically? Where can Japan obtain the energy imports that it will need?

Most of this chapter will deal with possible developments in the 1980s. A final section, however, will speculate about developments in the longer term. Unless otherwise indicated, all years referred to in this chapter are Japanese fiscal years.

Future Final Energy Consumption

Requirements for final energy consumption could be met in many ways. Over the long run, however, the combination of fuels actually

122

Table 7-1. *Structure of Final Energy Consumption in Japan by Principal Fuels, 1973, 1976, 1979, 1980, and Estimates for 1990*[a]
Percent

Fuel	Actual (calendar years)				Estimate, 1990	
	1973	1976	1979	1980	IEE	Official
Coal[b]	15.6	15.4	13.8	16.6	17.7	23.7[c]
Oil	67.6	65.8	66.3	62.5	58.7[d]	54.8
Gas	2.9	3.6	3.5	3.9	4.4	5.0
Other	0.3	0.2	0.1	0.2	. . .	. . .
Electricity	13.5	15.0	16.2	16.8	19.2	16.5
Total	100.0	100.0	100.0	100.0	100.0	100.0

Sources: Actual data are calculated from figures of total energy consumption in energy balance tables provided by the Institute of Energy Economics (Japan). IEE estimates are from IEE, paper (in Japanese) presented at an energy symposium, Tokyo, December 1981; calculations prepared May 1981. Official estimates are from Organization for Economic Cooperation and Development, International Energy Agency, *Energy Policies and Programmes of IEA Countries, 1980 Review* (Paris: OECD, 1981), pp. 190–91. Numbers are rounded.

a. Unless otherwise indicated, all references to years in this and subsequent tables in this chapter are to Japanese fiscal years, which begin on April 1 of the year specified.

b. Coal and other fuels (solar energy, liquefied and gasified coal) have been combined because the IEE does not show other fuels separately for all sectors.

c. Includes 8.1 percent use of other fuels.

d. The IEE does not give figures for nonenergy uses of fuels; therefore, estimates of nonenergy uses from the central case projections were added to the IEE oil estimates to make these figures comparable with those of the official estimates.

used will tend to reflect the economic interests of consumers. That is, for each kind of energy use, consumers will try to ensure that the marginal cost of useful energy is the same for all fuels. (If marginal costs differ, it is advantageous to use more of one fuel and less of another.)

The optimal combination of fuels is not easily achieved. The cost of replacing equipment designed to use a particular fuel retards reactions to change in the relative prices of different fuels. Uncertainty over future fuel prices also inhibits investments in new energy-using equipment. In recent years, moreover, governmental policies have strongly influenced choices of fuels.

The increase in Japan's dependence on oil before the 1973–74 oil crisis, and the effort to reduce that dependence since the crisis, were described in chapter 2. Japan has tried not only to check the overall rate of increase in energy consumption but also to substitute other fuels for oil. The results of this effort are evident in the changing shares of various fuels in total final energy consumption that are shown in the first four columns of table 7-1. From 1973 to 1979, success in reducing the share of oil in total final energy consumption was limited. A dramatic shift occurred in 1980, when the share of oil fell 3.8 percentage points. Most of this decline was made possible by an increase in the share of coal, although the shares of electricity and gas also increased.

At the time of the first oil crisis, coal prices rose with oil prices. There was therefore little incentive to switch from oil to coal. In the second oil crisis, however, oil prices rose sharply, but coal prices remained almost level. As a result, the iron and steel industry almost entirely gave up the use of heavy fuel oil in blast furnaces, and the cement industry shifted almost totally from oil to coal. Other less massive shifts have taken place in other industries.[1]

In estimating the future structure of final energy consumption, the key questions are whether the share of oil will decline and, if it does, what fuels will substitute for oil. Two recent estimates of final energy consumption in 1990 (last two columns of table 7-1) give somewhat different answers to these questions. One set of estimates, prepared in May 1981, is by the Institute of Energy Economics (Japan; IEE). The other set of estimates was prepared by the Japanese government in 1980 and was published by the Organization for Economic Cooperation and Development in 1981.

The differences between the two estimates of the structure of final energy consumption in 1990 diverge significantly. The share of oil in the official estimates is 3.9 percentage points lower than the share of oil in the IEE estimates, and the share of coal and other fuels is 6.0 percentage points higher. The share of electricity in the official estimates is 2.7 percentage points below its share in the IEE estimates.

The large difference in the share of coal and other fuels in the two estimates for 1990 appears to be the result of official optimism about the development of new energy sources, including solar energy and liquefied and gasified coal. The share of other fuels in the official estimates increases from 1.4 percent of total final energy consumption in 1985 to 8.1 percent in 1990.[2] The IEE estimate does not show other fuels separately (except for a very small amount allocated to the transportation sector in 1990) and makes little or no allowance for new energy sources. The increase in energy from new sources projected in the official

1. Institute of Energy Economics (Japan), *Energy in Japan*, no. 56 (March 1982), pp. 8–10.
2. Organization for Economic Cooperation and Development, International Energy Agency, *Energy Policies and Programmes of IEA Countries, 1980 Review* (Paris: OECD, 1981), pp. 190–91.

Table 7-2. *Final Energy Consumption in Japan by Fuels, 1980 and Estimates for 1990, Central Case*
Ten trillions of kilocalories[a]

Sector	Coal and other[b] 1980	1990	Oil[c] 1980	1990	Gas[d] 1980	1990	Electricity 1980	1990	Total 1980	1990
Industrial	42.9	57.2	68.4	87.2	2.3	3.6	26.9	40.8	140.4	188.8
Residential and commercial	1.8	3.5	29.0	40.0	8.0	11.9	15.9	24.6	54.7	80.0
Transportation	0	1.2	56.2	94.2	0	0	1.3	2.5	57.5	97.9
Nonenergy uses of fuels[e]	0	0	6.5	9.4	0	0	0	0	6.5	9.4
Total	44.7	61.9	160.1	230.8	10.3	15.5	44.1	67.9	259.2	376.1

Sources: Figures for 1980 are from energy balance tables provided by the IEE; totals for 1990 are the central case projections described in chapter 6; consumption of various fuels was calculated by using the percentage distribution of fuel consumption in each sector in recent IEE estimates. Numbers are rounded.

a. Ten trillion kilocalories approximately equal a million metric tons of oil equivalent (t.o.e.).
b. "Coal and other" includes coal products (coke, briquets, coke oven gas, and blast furnace gas) plus wood and charcoal.
c. "Oil" includes all refined petroleum products.
d. "Gas" includes natural gas, liquefied natural gas, and town gas.
e. In Japan, exclusively oil.

estimates also explains the lower shares of oil and electricity than those projected in the IEE estimate. Past experience justifies a conservative view toward the speed of developing new sources of energy.[3] The IEE estimate, rather than the official estimate, would therefore seem a better guide in determining what combination of fuels might be used to satisfy future energy consumption requirements.

Table 7-2 applies the IEE projections of fuel shares to the central case definition of energy requirements by sector in 1990 and derives estimates of the consumption of various fuels in those years (1980 figures are shown for purposes of comparison).[4] Total final energy consumption is projected in the central case to increase at an annual rate of 3.8 percent from 1980 to 1990. Electricity and gas increase at annual rates of 4.4

3. That most of the increase in the use of other fuels called for in the official estimates is in the industrial sector supports this point of view. The official figures suggest optimism that may not be warranted concerning the early production of liquefied and gasified coal.

4. The percentage shares of various fuels in total final energy consumption in table 7-2 and in the IEE projections diverge in 1990 because of differences in the sectoral distribution of energy consumption. In particular, the share of transportation in total final energy consumption is higher in table 7-2 than in the IEE projections (26.0 percent compared with 20.4 percent). Because oil is the dominant fuel in the transportation sector, the two estimates also differ with respect to the share of oil in total final energy consumption (61.3 percent in table 7-2 and 58.7 percent in the IEE projections).

percent and 4.2 percent, respectively. Oil increases 3.7 percent a year, and coal increases 3.3 percent a year. A different picture emerges, however, if transformation losses (that is, the use of primary fuels to make electricity and other secondary fuels) are taken into account.

Transformation Losses

Most of the fuels entering into final energy consumption are secondary products. The net losses of primary energy incurred in making these secondary products, known as transformation losses, must be added to final energy consumption in calculating total requirements for primary energy supplies. The most important transformation activities are the generation of electricity and the refining of petroleum. The energy used in making town gas and various coal products must also be accounted for.[5]

Electricity

The energy balance table for 1980 shows that a gross input of 135.3 million metric tons of oil equivalent (t.o.e) of primary energy was needed to produce electricity with an energy content of 50.0 million t.o.e.[6] Generating efficiency was therefore approximately 37 percent, and transformation losses were 85.3 million t.o.e. Some of these losses, however, were not real. As is the established practice, hydroelectric and nuclear power plants were assigned hypothetical energy inputs equal to the fuel that conventional thermal power plants would have required to generate the same amounts of energy. In estimating future transformation losses, it is therefore desirable to distinguish electricity generated by hydropower and nuclear power (and also by geothermal power as it is developed) from electricity generated by conventional thermal means.

In 1980 hydropower plants produced 466 thousand t.o.e. of electricity per gigawatt of capacity, and nuclear power plants produced 475 thou-

5. The transformation sectors of energy balance tables include the statistical difference needed to make the tables balance. These differences are not large in Japanese energy statistics and have been ignored in estimating transformation losses in 1990.

6. Energy balance table provided by the IEE. The amount of electricity delivered to final consumers was lower (44.1 million t.o.e) because some electricity was consumed within the transformation sector and lost in the process of transmission and distribution to consumers. Electricity is given a caloric value of 860 kilocalories per kilowatt hour.

Table 7-3. *Estimated Primary Electricity Generation in Japan, 1990*

Source	Generating capacity (gigawatts)	Estimated output (ten trillions of kilocalories)[a]
Hydropower	23.00	10,350
Nuclear power	36.00	18,000
Geothermal power	1.00	450
Total	60.00	28,800

Sources: Estimates of generating capacity are from Takao Tomitate, "Energy Supply Options for the 1980s: A Japanese Perspective," paper prepared for international energy conference, Tokyo (IEE, Research Division, May 1981), table 3. Estimates of output are explained in the text. Numbers are rounded.

a. Ten trillion kilocalories approximately equal a million t.o.e.

sand t.o.e. per gigawatt of capacity.[7] For purposes of projecting future generation of electricity, these relationships will be rounded off to 450 thousand t.o.e. per gigawatt of capacity in the case of hydropower and 500 thousand t.o.e. per gigawatt of capacity in the case of nuclear power. Output of hydropower per unit of capacity has been reduced slightly to allow for the possibility of water shortages. Output of nuclear power per unit of capacity has been raised slightly to reflect a possible increase in the percentage use of nuclear generating capacity.

IEE has estimated the generating capacities for hydroelectric, nuclear, and geothermal power in 1990 that are shown in the first column of table 7-3. If the assumptions stated above concerning electricity generation per gigawatt of capacity are used (and geothermal power is arbitrarily treated the same as hydropower), the estimates of primary electricity production shown in the second column of table 7-3 can be derived.

Conventional thermal power plants must make up the difference between projections of total electricity production and the sum of the estimates for hydroelectric, nuclear, and geothermal power (see totals in the fourth and fifth columns of table 7-4).[8] The projections of total

7. The energy balance table previously cited shows hypothetical energy inputs of 22,366 thousand t.o.e. for hydropower and 20,092 thousand t.o.e. for nuclear power in 1980. Using the generating efficiency of 37 percent, hydropower would have generated 8,275 thousand t.o.e. of electricity and nuclear power 7,434 thousand t.o.e. The generating capacities of hydropower and nuclear power were 17.76 gigawatts and 15.51 gigawatts, respectively. See IEE, *Energy in Japan*, no. 53 (June 1981), p. 11.

8. Treating conventional thermal power as the residual is generally consistent with its role in the overall electricity supply system. For reasons of efficiency, nuclear power plants are operated at a fairly constant rate to provide base-load power. Hydropower plants are operated at a more variable rate, but their total output is limited by the availability of water. Conventional thermal plants make up the difference in both base-load and peak-load power.

Table 7-4. *Structure of Thermal Electricity Generation in Japan, 1980 and Estimates for 1990*

Fuel	Composition of generating capacity (percent) 1980	1990	Estimated output, 1990 (ten billions of kilocalories)[a]	Estimated input, 1990 (ten billions of kilocalories)[a]
Coal	6.4	20.9	9,591	25.9
LNG[b]	24.1	34.1	15,648	42.3
LPG[c]	0.8	3.5	1,606	4.3
Other gas[d]	2.8	1.9	872	2.4
Oil[e]	66.1	39.6	18,172	49.1
Total	100.0	100.0	45,889	124.0

Source: Composition of generating capacity is taken from Hironobu Sato, "Long-Term Electric Power Plan for F.Y. 1981," in IEE, *Energy in Japan*, no. 53 (June 1981), p. 11. Numbers are rounded.
  a. Ten billion kilocalories approximately equal a thousand t.o.e.
  b. Liquefied natural gas.
  c. Liquefied petroleum gas.
  d. Includes coke oven gas, blast furnace gas, and domestic natural gas.
  e. Includes crude oil and refined petroleum products (except liquefied petroleum gas).

electricity production used here are the estimates of final electricity consumption from table 7-2 inflated by 10 percent to take account of both electricity consumption within the transformation sector and losses in transmission and distribution to consumers.[9]

In April 1981 the Central Electric Power Council published the annual revision of its Long-term Electric Power Plan. Among other features, this plan called for the striking shift in the composition of conventional thermal generating capacity shown in the first two columns of table 7-4. If electricity output is assumed to be proportional to capacity,[10] the electricity generated by conventional thermal plants in 1990 would be divided as shown in the third column of table 7-4. The inputs of various fuels required to produce the amounts of energy shown can be very roughly estimated (by assuming that all conventional thermal plants have a generating efficiency of 37 percent) as shown in the last column of table 7-4.[11]

9. In 1980 about 6 percent of electricity generated was lost in transmission and distribution, and about 4 percent was consumed within the transformation sector by electric power plants, town gas producers, refineries, and coal mines.
10. This assumption may, of course, be wrong. The newer plants that burn coal and liquefied natural gas may be able to produce electricity more cheaply than older plants that burn oil and may therefore be operated at higher capacity.
11. This assumption is more realistic for 1990 than it would have been in the past, when gas and oil clearly had higher thermal efficiencies than coal. The efficiency of new coal-burning power plants reportedly approaches that of oil-burning plants.

Table 7-5. *Estimated Inputs of Primary and Secondary Fuels in Thermal Electricity Generation in Japan, 1990*
Ten trillions of kilocalories[a]

Fuel	Estimated input
Primary	
Coal	25.9
Natural gas[b]	42.5
Crude oil[c]	13.4
Secondary	
Coal products[d]	2.2
Petroleum products[e]	40.0
Total	124.0

Source: Table 7-4. Numbers are rounded.
a. Ten trillion kilocalories approximately equal a million t.o.e.
b. Liquefied natural gas and domestic natural gas, assumed to be 10 percent of other gases (about the proportion in 1980).
c. One-fourth of the sum of liquefied petroleum gas and oil (again, about the proportion in 1980).
d. Other gases (coke oven gas and blast furnace gas) less domestic natural gas.
e. Oil and liquefied petroleum gas less crude oil.

To facilitate consideration of other parts of the transformation sector, the inputs shown in table 7-4 can be divided into primary and secondary fuels as in table 7-5. Transformation losses incurred in operating conventional thermal power plants are the difference between required fuel and electricity produced. These losses in 1990, in ten trillions of kilocalories (or millions of t.o.e.), can be summarized as follows: coal and coal products, 17.7; crude oil and petroleum products, 33.6; natural gas, 26.8; for a total of 78.1.

Petroleum Products

A small amount of energy is lost in the refining process, and refineries use some petroleum products in their own operations. As a result, the energy value of petroleum products turned out by refineries is about 94 percent of the energy value of the crude oil used in making those products. To make the same point in another way, the crude oil required to make a given volume of refined petroleum products available to final consumers has an energy value 1.064 times the energy value of those products.[12]

All of the oil entering directly into final consumption and most of the

12. The energy balance tables prepared by the IEE show transformation losses in petroleum refining of 5.8 percent in 1979 and 6.2 percent in 1980.

oil used to generate electricity is consumed in the form of refined petroleum products. In addition, a small amount of petroleum products is used in the manufacture of town gas.[13] Total requirements of petroleum products in 1990 can be estimated, in millions of t.o.e., as follows: direct consumption, 230.8; electricity generation, 40.1; town gas manufacture, 5.4; for a total of 276.3. The direct consumption figures are from table 7-2. The estimates of petroleum products used in the generation of electricity are explained in the immediately preceding section of this chapter. The figures for town gas manufacture are 35 percent of the estimates of direct gas consumption in table 7-2.

If all of the above estimated requirements for petroleum products were met by Japanese refineries, crude oil with a caloric value of 294.0 million t.o.e. would be required in 1990. Transformation losses in petroleum refining would then be 17.7 million t.o.e. in 1990.

If Japan continues to import some refined petroleum products, transformation losses will be correspondingly reduced. In 1979 and 1980 net imports of petroleum products (after stock changes) were 11.0 percent and 11.8 percent, respectively, of the petroleum products entering into final energy consumption. Predicting future imports of petroleum products with any precision is not possible; to allow for their possible effect, however, estimated transformation losses will be arbitrarily reduced by 10 percent to 15.9 million t.o.e. in 1990.

Coal Products

Most coal enters into final energy consumption as coke and related products, but the proportion of coal consumed directly is increasing. Important changes are taking place in the use of coal, as indicated in table 7-6, which shows the percentage shares of coal and coal products (plus wood and charcoal) in final energy consumption in calendar year 1973 and in fiscal years 1979 and 1980.

Coal, wood, and charcoal are of course part of primary energy supply. The other fuels listed in table 7-6 are secondary products made from coal at the cost of some transformation losses. Coke and coke oven gas are joint products, and blast furnace gas is a by-product of using coke to make steel. In 1973 these closely related products accounted for 92.2

---

13. In both 1979 and 1980, this equaled about 35 percent in caloric value of direct gas consumption.

Table 7-6. *Share of Coal and Coal Products in Final Energy Consumption in Japan, 1973, 1979, and 1980*
Percent

Item	1973[a]	1979	1980
Coal	4.4	7.8	14.4
Coke	65.0	66.6	63.5
Coke oven gas	12.4	13.5	12.0
Blast furnace gas	14.8	10.7	9.1
Briquets	1.7	0.7	0.5
Wood and charcoal	1.7	0.6	0.5
Total	100.0	100.0	100.0

Source: Energy balance tables provided by the IEE. Numbers are rounded.
a. Calendar year.

percent of total final consumption of coal, coal products, wood, and charcoal. In 1979 they accounted for 90.8 percent, but in 1980 their share fell to 84.6 percent.

Most production and use of coke, coke oven gas, and blast furnace gas is closely related to the production of steel. A tight quantitative relationship, however, between the output of steel and these fuels is not to be expected. Coke requirements of the steel industry vary inversely with the amount of scrap (as opposed to pig iron) used in making steel. Some coke, coke oven gas, and blast furnace gas are also used in generating electricity and in making town gas.

Nevertheless, the future level of steel production will be a major determinant of the level of output of these three fuels. Crude steel production was 113.0 million metric tons in 1979. It has been projected to increase to 130.6 million tons in 1990.[14] The total output of coke and related fuels was 42.5 million t.o.e. in 1979. If output of these fuels kept pace with steel production, it would be 49.1 million t.o.e. in 1990.[15]

In 1979, 46.5 million t.o.e. of energy in the form of coal were used in making coke and related fuels. If the relationship between coal inputs and gross outputs of these fuels remained the same, coal requirements for this purpose would be 53.7 million t.o.e. in 1990.

Transformation losses in making coke and related fuels are the difference between coal consumption and gross output plus the use of

14. IEE projection, May 1981.
15. This assumption may slightly understate future coke requirements if the amount of energy required to produce a metric ton of steel rises 0.2 percent annually in the 1980s, as is projected in table 6-1.

these fuels by their producers. These losses were 3.6 percent of the gross output in both 1979 and 1980. Transformation losses in making these fuels may therefore be estimated at 6.5 million t.o.e. for 1990. This estimate should be increased by perhaps 0.3 million t.o.e. to allow for the consumption of coal by coal mines.[16] These estimates of coal requirements and transformation losses take account of coal products used to make electricity and town gas, as well as those consumed in the steel industry and elsewhere.[17]

Town Gas

Town gas is a mixture of natural gas (both domestic natural gas and imported liquefied natural gas) and gas from petroleum products (naphtha, liquefied petroleum gas, and refinery gas), and coal products (coke, coke oven gas, and blast furnace gas). Transformation losses in making the petroleum and coal products in town gas have already been taken into account. Additional small losses, however, are incurred in making town gas, and some town gas is used by town gas producers.

In 1980, 91.7 percent or 9.4 million t.o.e. of the gas entering into final gas consumption was town gas. The remainder was domestic natural gas and liquefied natural gas distributed to final consumers (mostly large companies) without being mixed with the other components of town gas. Transformation losses, including self-use by town gas producers, were 0.6 million t.o.e. or about 6 percent of final consumption.

The IEE recently estimated that town gas consumption would be 14.2 million t.o.e. in 1990. At a rate of 6 percent, transformation losses would then be roughly 0.8 million t.o.e. in 1990. These losses could take the form of a mixture of the various components of town gas. In the interest of simplicity, however, they will be assumed to consist entirely of natural gas.[18]

In summary, total transformation losses in 1990 (including the hypothetical losses attributed to the generation of electricity by hydroelectric,

16. The actual coal consumption by coal mines in 1979 and 1980 was 311 thousand t.o.e and 245 thousand t.o.e., respectively.

17. Transformation losses incurred in making coal briquets have become so small that no effort has been made to estimate their future size. Briquets therefore have in effect been included with coal, wood, and charcoal as untransformed fuels.

18. Natural gas (domestic and imported) made up 48.9 percent of town gas in 1979 and 50.5 percent in 1980.

Table 7-7. *Total Estimated Transformation Losses in Japan, 1990*
Ten trillions of kilocalories[a]

Fuel	Estimated loss
Coal	24.6
Oil	51.3
Gas	29.0
Subtotal	104.9
Hydropower	18.6
Nuclear power	32.2
Geothermal power and other	0.8
Subtotal	51.6
Total	156.5

Source: See the text and tables 7-3 through 7-6. Numbers are rounded.
a. Ten trillion kilocalories approximately equal a million t.o.e.

nuclear, and geothermal power) can be estimated as shown in table 7-7. In 1980 transformation losses were 30.8 percent of primary energy supplies (that is, final energy consumption plus transformation losses). In the central case these losses are projected to be 29.5 percent of primary energy supplies in 1990.

Primary Energy Supplies

Table 7-8 draws on the preceding analyses of final energy consumption and transformation losses to present the composition of Japan's actual primary energy supplies in 1980 and its estimated requirements for primary energy in 1990. Total requirements for primary energy are projected to increase at an average annual rate of 3.6 percent from 1980 to 1990. Given the 5.0 percent annual growth in gross domestic product assumed in the central case, the average income elasticity of demand for energy projected for this period is 0.72.

Although this illustrative projection was developed for the period 1980–90, it might be slipped forward to apply to 1982–92 or 1983–93. Japan's primary energy supplies in 1982 were 6.6 percent smaller than in 1980, even though its real GDP was 6.1 percent greater. Primary energy supplies probably declined further in 1983.[19]

19. Toshiaki Yuasa, "Has Energy Demand Bottomed in Japan?—Outlook from Trend in Japan," IEE, *Energy in Japan*, no. 68 (March 1984), pp. 1–7. The unexpected opposite movements of GDP and primary energy supplies suggest important structural changes in the Japanese economy.

Table 7-8. *Primary Energy Supplies in Japan by Fuel, 1980,*
and Estimated Requirements for Primary Energy, 1990
Ten trillions of kilocalories[a]

Fuel	Actual supplies, 1980	Estimated require- ments, 1990
Coal[b]		
Consumed as coal	6.7	19.1
Consumed as coal products[c]	42.3	51.3
Used to generate electricity	11.7	28.1
Statistical difference	2.9	. . .
Subtotal	63.6	98.5
Oil[d]		
Consumed as petroleum products	176.8	249.8
Used to generate electricity[e]	62.7	55.7
Statistical difference	4.3	. . .
Subtotal	243.8	305.5
Gas[f]		
Consumed as gas	10.7	16.3
Used to generate electricity	18.0	42.5
Coal and petroleum products used in town gas[g]	−5.1	−7.1
Statistical difference	0.7	. . .
Subtotal	24.3	51.7
Nuclear power	20.1	48.6
Hydropower	22.4	28.0
Geothermal power and other	0.2	1.2
Total	374.4	533.5

Sources: Figures for 1980 are from energy balance tables prepared by the IEE. For 1990 estimates, see the text. Numbers are rounded.

a. Ten trillion kilocalories approximately equal a million t.o.e.
b. Includes wood and charcoal.
c. Includes coal consumed by coal mines and coal products used to make town gas.
d. Includes all refined petroleum products and crude oil used to generate electricity (including imported petroleum products and petroleum products used to make town gas).
e. Includes crude oil burned by power plants.
f. Includes domestic natural gas, liquefied natural gas, and town gas (including that used by town gas producers).
g. Subtracted to avoid double counting.

In the central case estimates (table 7-8), the various kinds of primary energy are projected to grow at the following annual percentage rates from 1980 to 1990: coal, 4.5; oil, 2.3; gas, 7.8; nuclear, 9.2; hydroelectric and other, 2.6. The low rate of increase projected for oil would represent substantial success for Japan's efforts to shift to other sources of energy. Table 7-9 shows how the different rates of growth projected for various kinds of primary energy would affect shares of primary energy supply.

Table 7-9. *Structure of Primary Energy Supply in Japan, 1980, and Projected Structure, 1990*
Percent

Fuel	1980	1990
Coal	17.0	18.4
Oil	65.1	57.3
Gas	6.5	9.7
Nuclear power	5.4	9.1
Hydropower and other	6.0	5.5
Total	100.0	100.0

Source: Based on table 7-8. Numbers are rounded.

Table 7-8 also suggests that significant changes may take place in the ways in which primary energy is used. These changes in use, based on table 7-8, are summarized below (in percent) for 1980 and 1990:[20]

	1980	1990
Consumed directly	3.4	5.3
Used to make secondary products	59.8	56.4
Used to generate electricity	36.9	38.2

Although the figures above do not project any radical changes in Japan's energy economy, they do suggest that more energy may be consumed directly as coal and gas and as electricity, and that less may be consumed as secondary coal and petroleum products.

Import Requirements

In the absence of major discoveries of domestic energy resources—which are not expected—Japan must continue to import most of the hydrocarbons that it consumes. According to IEE energy balances, in 1980 Japan imported 81.6 percent of its coal, 90.5 percent of its natural gas, and 99.8 percent of its oil. Recent forecasts of domestic energy production, taken in conjunction with the estimates of primary energy requirements in table 7-8, suggest that by 1990 dependence on imported coal may rise to 87.0 percent, but that dependence on imported gas may

20. The small statistical difference in 1980 data was ignored in calculating percentages for that year. Coal products and petroleum products used in town gas were subtracted before calculating the percentages of primary energy supplies consumed directly. Numbers are rounded.

Table 7-10. *Japan's Dependence on Imported Hydrocarbons, 1980,*
and Estimated Dependence, 1990
Ten trillions of kilocalories[a]

Fuel	1980	1990
Coal		
Domestic	11.7	12.8
Imported	51.9	85.7
Total	63.6	98.5
Percent imported	81.6	87.0
Oil		
Domestic	0.4	1.7
Imported	243.4	303.8
Total	243.8	305.5
Percent imported	99.8	99.4
Gas		
Domestic	2.3	6.4
Imported	22.0	45.3
Total	24.3	51.7
Percent imported	90.5	87.6
All hydrocarbons		
Domestic	14.4	20.9
Imported	317.3	434.8
Total	331.7	455.7
Percent imported	95.6	95.4

Sources: Figures for 1980 are from energy balance tables provided by the IEE. Estimates of domestic production are from OECD, IEA, *Energy Policies and Programmes, 1980*, p. 190. Total requirements for each fuel are from table 7-8. Imports are the difference between total requirements and domestic production. Numbers are rounded.
a. Ten trillion kilocalories approximately equal a million t.o.e.

decline to 87.6 percent. Dependence on imported oil may decline marginally to 99.4 percent. The percentage of all hydrocarbons that must be imported would remain about the same in 1990 as it was in 1980: 95.4 percent compared with 95.6 percent (table 7-10).

Hydrocarbon imports would be reduced, however, from 84.7 percent of all primary energy supplies in 1980 to 81.5 percent in 1990. Of particular significance, oil imports would also fall—from 65.0 percent of primary energy supplies in 1980 to 56.9 percent in 1990. Oil imports would nevertheless be about 25 percent greater in 1990 than in 1980, increasing from 4.9 million to 6.1 million barrels a day.

Japan's uranium reserves are so small that it can be assumed that all of the uranium required by its nuclear power plants must be imported.[21]

21. Uranium reserves are estimated to be only 10,000 tons of U_3O_8. See Power Reactor and Nuclear Fuel Development Corporation, *PNC* (a brochure describing the programs of the PNC) (Tokyo: PNC, n.d.), p. 4.

Table 7-11. *Estimated Japanese Energy Imports, 1990, Central Case*

Fuel	1990
Coal (millions of metric tons)	123.6
Oil (millions of metric tons)	303.8
Gas (billions of cubic meters)	46.2
Uranium (thousands of short tons of U_3O_8)	8.3

Sources: Estimated oil import requirements are from table 7-10. The following ratios were used to convert coal and gas import requirements to physical units: imported steam coal, 6,500 kilocalories per kilogram; imported coking coal, 7,200 kilocalories per kilogram; natural gas, 9,800 kilocalories per cubic meter. Steam coal was assumed to constitute 38 percent of coal imports (by weight) in 1990. Conversion ratios were provided by the IEE. Percentages of steam coal in total coal imports were derived from Tomitate, "Energy Supply Options," table 3. Numbers are rounded.

In 1978 an official Japanese government committee estimated annual uranium requirements at 17,300 short tons of U_3O_8 in 1990.[22] Nuclear generating capacity, however, was assumed to be 60 gigawatts (electric) [GW(e)] in 1990, compared with the estimate of 36 GW(e) in 1990 used in this study. If the committee's estimates are scaled down to these lower capacities, annual uranium requirements become about 10,400 short tons in 1990. If spent nuclear fuel were to be reprocessed and the residual uranium recycled, import requirements would be correspondingly reduced. The same government committee assumed that 2,100 tons of residual uranium would be recycled in 1990. If this assumption is accepted, Japan's requirements for imported uranium would be 8,300 tons in 1990.[23]

Under the assumptions of the central case, Japan's energy imports in 1985 and 1990 in original physical units can be estimated as in table 7-11. These import requirements would of course be somewhat different if the annual rate of growth of GDP were higher or lower than 5.0 percent or if real energy prices were not to remain stable. Changing the assumptions of the central case about GDP growth or energy prices would not, however, increase or decrease imports of all fuels in the same proportion. The differential effects of changed assumptions can be illustrated by

22. Ministry of International Trade and Industry (Japan), Research Committee on Nuclear Fuel, "Interim Report on the Results of Study on Nuclear Fuel Cycle," summarized in *Atoms in Japan*, vol. 22 (September 1978), pp. 27–34.

23. These figures appear to include a margin above actual consumption that could be added to stocks. Under the rule of thumb that each 1,000 megawatts (electric) [MW(e)] of capacity requires 142 metric tons of uranium or 184.6 short tons of U_3O_8 annually, requirements would be 6,600 short tons of U_3O_8 in 1990.

Table 7-12. *Estimated Contribution of Various Fuels*
to Final Energy Consumption in Japan in 1990,
Central Case and Alternatives
Ten trillions of kilocalories[a]

	Annual GDP growth rate (percent)		
Fuel	4	5[b]	6
Coal and other	59.0	61.9	64.9
Oil	214.5	230.8	248.3
Gas	14.5	15.5	16.6
Electricity	64.2	67.9	71.8
Total	352.2	376.1	401.6

Sources: Estimates were calculated from the sectoral energy consumption estimates in table 6-18 and from the shares of various fuels in sectoral consumption in IEE projections. Numbers are rounded.
a. Ten trillion kilocalories approximately equal a million t.o.e.
b. Central case.

estimating import requirements under two alternatives to the central case: 4.0 percent and 6.0 percent annual growth of real GDP.[24]

Table 7-12 compares final energy consumption in the central case (see table 7-2) with alternative cases in which real GDP is assumed to increase 4.0 percent and 6.0 percent annually. In all cases real energy prices are assumed to remain fairly stable. Sectoral shares of final energy consumption in each case were taken from table 6-18. The percentage distribution of consumption of various fuels within sectors was assumed to be unaffected by the GDP growth rate.[25]

A rough estimate of total requirements for primary energy under the alternatives to the central case can be obtained by assuming that primary energy inputs bear the same relationship to final energy outputs as they do in the central case.[26] This assumption produces the results shown in table 7-13.

The next question is how much of the various primary sources of

24. Alternatives to the central case that involve changes in real energy prices were not chosen because it has not been possible to calculate sectoral price elasticities of energy demand. If such elasticities were available, the procedures followed in estimating import requirements would be similar to those used in the alternative cases that are based on changes in the assumed GDP growth rate.

25. The composition of fuel consumption within sectors could be somewhat different under different rates of GDP growth, but available data do not provide an adequate basis for testing this possibility.

26. This assumption has the effect of holding constant the ratio of coal consumed as coal to coal consumed as coal products.

Table 7-13. *Estimated Total Japanese Requirements of Primary Energy in 1990, Alternatives to Central Case*
Ten trillions of kilocalories[a]

Fuel	Annual GDP growth rate (percent)	
	4	6
Coal	67.1	73.8
Oil	232.1	268.7
Gas	15.2	17.5
Electricity	193.0	215.8
Subtotal	507.4	575.8
Adjustment[b]	−6.8	−7.9
Total	500.6	567.9

Source: See the text. Numbers are rounded.
a. Ten trillion kilocalories approximately equal a million t.o.e.
b. Petroleum products and coal products used in town gas are subtracted to avoid double counting. Such products are assumed to be 45 percent of gas.

energy will be used to generate electricity in the alternatives to the central case. One plausible possibility is that the hypothetical energy inputs of nuclear, hydroelectric, and geothermal power will be the same as those in the central case. Because of the long lead times for planning and construction, total generating capacity of nuclear, hydroelectric, and geothermal power plants would probably not be much affected by the moderate differences in GDP growth under consideration here. Moreover, as pointed out earlier, the level of operation of such plants would be determined largely by factors other than total demand for electricity.

Given these assumptions about nuclear, hydroelectric, and geothermal power, the total inputs of primary energy required in 1990 for conventional thermal power plants under 4.0 percent and 6.0 percent GDP growth would be as shown in table 7-14 (total row, first two columns of the table). For purposes of comparison, the central case total requirements (under 5.0 percent GDP growth) would be 126.3 million t.o.e. in 1990. The first two columns of table 7-14 also show for the alternative cases the allocation of total requirements among various hydrocarbon fuels. The same proportions as in the central case have been used.[27]

27. The shares of various fuels may be different at different levels of requirements. The shift to fuels other than oil might be somewhat accelerated if requirements are high. If low requirements cause an increase in idle generating capacity, utility companies might rely more on coal-burning plants whose fuel is relatively cheap.

Table 7-14. *Estimated Japanese Requirements of Hydrocarbon*
Fuels for Thermal Power Generation and All Uses in 1990,
Alternatives to Central Case
Ten trillions of kilocalories[a]

	Requirements for thermal power generation		Total requirements	
	4	6	4	6
	percent	percent	percent	percent
	GDP	GDP	GDP	GDP
Fuel	growth	growth	growth	growth
Coal[b]	25.6	30.7	92.7	104.5
Oil[c]	50.8	60.8	282.9	329.5
Gas[d]	38.8	46.4	47.2	56.0
Total	115.2	138.0	422.8	490.0

Source: See the text. Numbers are rounded.
a. Ten trillion kilocalories approximately equal a million t.o.e.
b. Includes gas made from coal.
c. Includes liquefied petroleum gas.
d. The entire adjustment needed to avoid double counting of petroleum products and coal products used in town gas has been subtracted from gas requirements.

Estimates of total requirements for hydrocarbons under 4.0 percent and 6.0 percent GDP growth can be obtained by adding the estimated requirements for thermal generation of electricity to the earlier estimates (table 7-8) of requirements for other uses of hydrocarbons. The results are shown in the last two columns of table 7-14.

There is little reason to expect that domestic production of hydrocarbons will vary with the rate of GDP growth. If the same estimates of domestic production are used in all three cases (see table 7-10), the estimates of import requirements shown in table 7-15 are obtained. The table also shows these requirements in original physical units. The assumption that electricity generated in nuclear power plants would be the same in the alternative cases as in the central case means that uranium import requirements would also be the same as those shown in table 7-11 (that is, 8,300 short tons of U_3O_8 in 1990).

The most remarkable feature of the import requirements in the alternative cases is their small divergence from those in the central case (table 7-11). For example, oil import requirements in 1990 in the case of 4.0 percent GDP growth are only 7.4 percent below those in the central case; import requirements in the case of 6.0 percent GDP growth are only 7.9 percent above those in the central case.

These outcomes, of course, are the result of the assumptions under-

Table 7-15. *Estimated Japanese Energy Imports in 1990,*
Central Case and Alternatives

Fuel	Annual GDP growth rate (percent)		
	4	5[a]	6
Coal			
Ten trillions of kilocalories[b]	79.9	85.7	91.7
Millions of metric tons	115.2	123.6	132.2
Oil			
Ten trillions of kilocalories[b]	281.2	303.8	327.8
Millions of metric tons	281.2	303.8	327.8
Gas			
Ten trillions of kilocalories[b]	40.8	45.3	49.6
Billions of cubic meters	41.6	46.2	50.6
Total (ten trillions of kilocalories)[b]	401.9	434.8	469.1

Sources: Tables 7-10 and 7-11 (see the latter for ratios used to convert kilocalories to physical units and for assumption about the share of steam coal in total coal requirements). Numbers are rounded.
a. Central case.
b. Ten trillion kilocalories approximately equal a million t.o.e.

lying the estimates, some of which were unavoidably quite arbitrary. The assumptions most open to challenge are holding constant, in all cases of GDP growth, the sectoral distribution of fuels in final energy consumption and the shares of various fuels used by conventional thermal electric power plants.

Possible Sources of Energy Imports

There is no reason to assume that Japan will expand its imports of various fuels simply by increasing proportionately the quantities provided by different energy-exporting countries in the recent past. Prospects for the expansion of fuel exports from these countries depend on geological, economic, and political factors that vary widely. Moreover, Japan's continuing efforts to diversify its sources of energy imports will influence and change future import patterns.

Coal

Table 7-16 shows the origins of Japan's imports of coking coal and steam coal in calendar years 1978–81. In this brief four-year period

The Energy Balance in Northeast Asia

Table 7-16. *Origins of Japan's Coal Imports,*
Calendar Years 1978–81
Millions of metric tons

Type of coal and country	1978		1979		1980		1981	
	Quantity	Percent	Quantity	Percent	Quantity	Percent	Quantity	Percent
Coking coal								
Australia	24.5	47.0	26.0	44.4	25.8	37.8	29.1	37.2
United States	8.9	17.1	13.5	23.1	19.2	28.2	21.6	27.6
Canada	10.9	20.9	10.5	17.9	10.6	15.5	9.6	12.3
USSR	2.3	4.4	2.2	3.8	1.9	2.8	1.1	1.4
South Africa	2.4	4.6	2.2	3.8	2.9	4.2	3.0	3.8
China	0.3	0.6	0.7	1.2	1.0	1.5	1.2	1.5
Other	0.9	1.7	1.0	1.7	0.4	0.6	0.2	0.2
Subtotal	50.2	96.4	56.1	95.9	61.8	90.6	65.8	84.0
Steam coal[a]								
Australia	0.7	1.3	1.0	1.7	3.5	5.1	5.7	7.3
China	0.6	1.2	0.7	1.2	1.1	1.6	1.6	2.0
South Africa	0.1	0.2	0.1	0.2	0.3	0.4	1.3	1.7
USSR	0.1	0.2	0.1	0.2	0.2	0.3	0.3	0.4
Vietnam	0.4	0.8	0.3	0.5	0.3	0.4	0.2	0.2
Other	. . .	. . .	0.2	0.3	1.0	1.5	3.5	4.5
Subtotal	1.9	3.6	2.4	4.1	6.4	9.4	12.5	16.0
Total	52.1	100.0	58.5	100.0	68.2	100.0	78.3	100.0

Sources: Quantities in 1978 and 1979 are from Japan External Trade Organization, *White Paper on International Trade: Japan 1980* (Tokyo: JETRO, 1980), pp. 129–30. Quantities in 1980 and 1981 are from the 1981 and 1982 editions of the same publication, pp. 175–76 and pp. 187–88, respectively. Numbers are rounded.

a. Includes anthracite and general (or ordinary) coal.

important changes took place. Steam coal imports increased sixfold but were still only one-eighth of total coal imports in 1981. Australia was Japan's leading source of both coking coal and steam coal, but it lost considerable ground to the United States as a supplier of coking coal.

Three countries—Australia, the United States, and Canada—supplied nearly seven-eighths of Japan's coal imports in 1981. The country ranking fourth, South Africa, provided only 5.5 percent of total coal imports in 1981—over 8 percentage points behind Canada, which ranked third. Minor coal exporters in the period 1978–81 included China, the USSR, and Vietnam. The market share of China increased somewhat over these three years, but it was only 2.5 percent in 1980. The shares of the USSR and Vietnam declined.

Most of the growth in Japan's coal imports during the 1980s will be in steam coal, especially in the second half of the decade. Requirements for coking coal will be restrained by the anticipated slow growth of the iron and steel industry (see table 6-1), whereas requirements for steam coal will rise as more coal-burning electric power plants come on line.

As noted earlier, the projected level of coal imports in the central case assumes the following allocations, in millions of metric tons, of coking coal and steam coal in 1990 (see also table 7-11; actual figures for 1980 are shown for comparison):

	1980	1990
Coking coal	61.8	76.6
Steam coal	6.4	47.0
Total	68.2	123.6

Because of its physical resources and transportation facilities, Australia is in a good position to increase its share of Japan's coal imports. Australian coal reserves are large (about 100 billion tons of measured plus indicated reserves). Nearly half of these reserves is in bituminous coal that has relatively high heat content and low sulfur content, and about 40 percent of the bituminous reserves is of metallurgical quality. Most of the reserves are easily mined and are located within 100 to 225 railroad miles of an ocean port. Rail and port facilities are adequate and are expected to keep pace with rising export volume.[28]

Frequent labor disputes make Australia an unreliable supplier for coal-importing countries, including Japan.[29] The Japanese government would like to reduce dependence on Australian imported coal from the 44.5 percent recorded in 1981 to about 30 percent in 1990.[30] This goal might be attainable if only the modest increase projected for imports of coking coal were involved, but holding Japan's dependence on Australian steam coal to 30 percent would be difficult if not impossible.[31]

Reducing Japanese dependence on Australia for imports of coking coal in 1990 to 30 percent would require increasing imports from other countries by 16.9 million metric tons a year (a reduction of 6.1 million metric tons in the 1981 level of imports from Australia plus all of the projected increase of 10.8 million metric tons in total imports from 1981 to 1990). The most promising sources of additional coking coal appear to be Canada, the United States, and China.

28. U.S. Department of Energy, *Interim Report of the Interagency Coal Export Task Force*, draft for public comment, DOE/FE-0012 (Government Printing Office, January 1981), pp. 49–51. Also see Akira Chimura, "The Recent Situation of Coal Development," in IEE, *Energy in Japan*, no. 61 (January 1983), pp. 3–5.

29. U.S. Department of Energy, *Interim Report*, and conversations in Sydney and Tokyo (February 1982).

30. Conversation at the IEE (Tokyo, February 1982).

31. Alternatives to Australia as a source of imported coal are based on U.S. Department of Energy, *Interim Report*, and on conversations in Sydney and Tokyo (February 1982).

U.S. exports of coking coal to Japan have increased rapidly. Exports from Canada leveled off in the late 1970s, but Canadian reserves of coking coal are well situated (in the western provinces) to supply the Japanese market. Some increases in exports of coking coal from China to Japan appear likely, although inadequate Chinese rail and port facilities will limit the extent of the increase.[32] An increase in exports of coking coal to Japan from the USSR and South Africa is also possible. Soviet exports, however, would be limited by deficient infrastructure, a deficiency possibly compounded by political problems,[33] and only a small fraction of South Africa's coal reserves is of metallurgical quality.

In 1981 Australia had 45.6 percent of Japan's market for imported steam coal. It is difficult to envision Japan's reducing this market share to 30 percent by 1990 while simultaneously increasing its imports of steam coal almost fourfold, from the 12.5 million metric tons imported in 1981 to the 47.0 million tons of imports projected for 1990. The main obstacle to a drastic reduction in Australia's market share is economic. Largely because of lower inland transportation costs, Australian steam coal could be delivered to Japan in 1980 for $1.60 (1980 U.S. dollars) per million British thermal units (Btus), compared with $2.25 for both U.S. and Canadian steam coal and $1.95 for South African steam coal.[34] Comparable cost figures for other suppliers are not available.

Chinese coal could compete with Australian coal because of its low inland and ocean transportation costs. Soviet coal from Siberia also has the advantage of a short haul by sea, but the mines are located about 1,500 miles from the port. Chinese exports of steam coal to Japan will be hampered by production and infrastructural deficiencies, although these should gradually be ameliorated by a substantial Japanese invest-

32. Japanese steel companies bought 1.2 million metric tons of Chinese coking coal in 1981 and contracted to buy 2.0 million metric tons in 1982. See *Japan Economic Journal*, June 12, 1982, p. 6.

33. Controversy over Soviet refusal to give up the Northern Territories (the southern Kuril Islands and several small islands closer to Hokkaido) that were occupied at the end of World War II continues to exert a chilling influence on relations between Tokyo and Moscow. Also, Japan would not want to appear to be moving closer to the Soviet Union in a period of worsening U.S.-Soviet relations. Yet these political complications have not thus far eliminated the continuing Japanese interest in sharing in the development of the natural resources of Siberia. For a good discussion of the factors influencing Japanese behavior in this regard, see Allen S. Whiting, *Siberian Development and East Asia: Threat or Promise?* (Stanford University Press, 1981), especially pp. 112–59.

34. U.S. Department of Energy, *Interim Report*, p. 99.

ment program.[35] The Japanese have thus far been more interested in obtaining coking coal than steam coal from Siberia. As noted above, political differences could also inhibit any large increase in Japanese imports of Soviet coal.

The major uncertainty in predicting Japan's future imports of steam coal is how large a premium Japanese utilities are willing to pay for the greater reliability of South African, Canadian, and U.S. coal. Per Btu, South African steam coal cost 22 percent more than Australian coal in 1980, and Canadian and U.S. steam coal was 41 percent more expensive than Australian coal. That Japan obtained over 10 percent of its steam coal from South Africa in 1981 suggests that the premium on South African coal was not prohibitive. U.S. and Canadian exports of steam coal to Japan in 1980 were too small to be listed separately in the statistics, but in 1981 they were 16.8 percent and 9.1 percent, respectively, of Japan's total imports of steam coal. Japanese utilities and trading firms have shown continuing interest in obtaining steam coal from the western United States (including Alaska) and western Canada.

The competitive advantage enjoyed by Australia in supplying steam coal to Japan could narrow because of rising labor costs and higher taxes. Moreover, the Japanese government could make it easier for utilities to pay a premium for non-Australian coal by approving higher electricity rates. Australia's continued dominance of the Japanese market for imported steam coal is therefore by no means assured.

Estimating the future geographical origins of Japan's coal imports (of both coking coal and steam coal) is clearly difficult. An analyst at the IEE provided the following rough percentage estimates of the origins of Japan's coal imports in 1990: Australia, 55–60; China and the USSR, 10; Canada, 10; United States, 15–20; South Africa and other, 10–15.[36] The most striking feature of these figures is the projected increase in Japan's dependence on Australian coal.

A recent Australian study estimates Australian coal exports to Japan in 1990 at 56 million metric tons (34 million tons of coking coal and 22 million tons of steam coal).[37] If Japan's total projected coal imports in 1990 are 123.6 million tons (table 7-11), Australia's market share would

35. See Chimura, "The Recent Situation of Coal Development."
36. Conversation in Tokyo (February 1982).
37. *Australian Economic Review*, cited in the *Sydney Morning Herald*, December 2, 1981.

Table 7-17. *Origins of Japan's Imports of Crude Oil,*
Calendar Years 1978–81
Millions of kiloliters

	1978		1979		1980		1981	
Country	Quantity	Percent	Quantity	Percent	Quantity	Percent	Quantity	Percent
Saudi Arabia	89.0	33.0	97.0	34.5	89.9	35.3	85.6	37.6
Indonesia	34.9	12.9	40.6	14.4	36.4	14.3	36.6	16.1
Iran	46.8	17.3	27.2	9.7	16.8	6.6	7.8	3.4
United Arab Emirates	27.4	10.1	28.6	10.2	34.5	13.6	31.2	13.7
Kuwait	23.2	8.6	27.1	9.6	10.1	4.0	10.6	4.7
Iraq	8.9	3.3	15.2	5.4	19.8	7.8	3.4	1.5
Oman	10.2	3.8	11.1	3.9	7.8	3.1	9.3	4.1
Brunei	8.7	3.2	9.5	3.4	8.0	3.1	5.2	2.3
China	8.7	3.2	8.5	3.0	9.1	3.6	10.3	4.5
Qatar	6.3	2.3	8.1	2.9	7.7	3.0	8.1	3.6
Malaysia	5.3	2.0	6.8	2.4	5.7	2.2	4.4	1.9
Other	1.3	0.5	1.5	0.5	8.6	3.4	14.9	6.6
Total	270.7	100.0	281.2	100.0	254.4	100.0	227.4	100.0

Sources: Same as table 7-16. Crude oil includes partly refined oil. Numbers are rounded.

be 45.3 percent, which is slightly higher than the 44.5 percent recorded in 1981.

Accepting an increase in Japan's dependence on Australian coal means assuming the complete failure of Japanese efforts to reduce that dependence. Assuming at least modest success may be more realistic. If Australia's share of Japan's coal imports were reduced to, say, 40 percent in 1990, relatively small increases in imports from other countries would be required. These increases would be most likely to come from South Africa, China, and, if the needed infrastructure to export coal from the West Coast is provided in time, the United States.[38]

Oil

Table 7-17 shows the geographic origins of Japan's imports of crude oil in calendar years 1978–81. Japan's oil imports rose 4.4 percent in 1979 but fell 19.1 percent in the following two years, largely as a result of an increase in the international price of oil. The share of various oil-exporting countries in the Japanese market also did not remain constant.

38. An early increase in exports to Japan of coal from the western United States was put in doubt in November 1982 by the postponement of plans to expand coal-loading facilities at the ports of Long Beach and Los Angeles. See *Japan Economic Journal*, November 16, 1982, p. 8.

As a result of domestic disorders and war with Iraq, Iran's share fell from 17.4 percent in 1978 to 3.4 percent in 1981. Despite the war, Iraq increased its share from 3.3 percent in 1978 to 7.8 percent in 1980, but its share fell to only 1.5 percent in 1981. The abrupt drop of Kuwait's share from 9.6 percent in 1979 to 4.0 percent in 1980 was associated with a decline of one-third in that country's total oil production.[39] The sharp rise in the share of the "other" category—from 0.5 percent in 1979, to 3.4 percent in 1980, and to 6.6 percent in 1981—is also noteworthy. Mexico was the largest supplier in this category.

In 1981 Saudi Arabia supplied 37.6 percent of Japan's total imports of crude oil and has continued to be Japan's largest supplier. Among other suppliers, only Indonesia (16.1 percent) and the United Arab Emirates (13.7 percent) provided more than 10 percent of Japan's crude oil requirements. About two-thirds of Japan's imports of crude oil in 1981 were from Arab countries. Nearly seven-tenths were from the Middle East.

Earlier in this chapter Japan's requirements for imported oil in 1990 were projected, under assumptions of the central case, to be 303.8 million metric tons, or 79.0 million tons more than the 224.8 million metric tons imported in 1980.[40] This increase in crude oil imports can probably be procured. (If it cannot, because of an excess of total demand over supply at prevailing international prices, prices would rise and all oil-importing countries, including Japan, would import less.) The important question is whether Japan can increase its oil imports by even this modest amount—about 1.3 million barrels a day—without simultaneously increasing its dependence on Middle Eastern or Arab oil or both.

In 1980, 68.6 percent of the Organization of Petroleum Exporting Countries' production capacity was in the Middle East; 69.9 percent of OPEC's capacity was in Arab countries. In 1990 the share of Middle Eastern countries in total OPEC capacity has been projected to increase to 73.7 percent; the share of Arab countries has been projected to remain about the same (69.8 percent), as in 1980.[41] The capacities of all OPEC

39. U.S. Department of Energy, *Monthly Energy Review*, February 1982, p. 90.

40. This figure, which is from an energy balance table supplied by the IEE, equals about 242.5 million kiloliters. It does not match exactly with the total imports for 1980 shown in table 7-18 because it is for the fiscal year, rather than the calendar year, and because it does not include imports of partly refined oil.

41. U.S. Department of Energy, Energy Information Administration, *1980 Annual Report to Congress*, vol. 3: *Forecasts*, table 2.3, p. 10. The 1990 percentages are based on the midrange projections.

members outside the Middle East were projected to decline or remain the same. Among the non-Arab members of OPEC, an increase in capacity was foreseen only for Iran (from 3.0 million barrels a day in 1980 to 4.0 million in 1990).

Iran may have some excess capacity in 1990 (estimated at 0.7 million barrels a day), but the non-Arab OPEC countries outside the Middle East are expected to produce at capacity. In contrast with this prospect, the Arab members of OPEC are projected to have 6.3 million barrels a day of excess capacity in 1990. All of this excess would be in countries bordering the Persian Gulf. (Libya and Algeria are projected to produce at capacity.) Over half of the excess, or 3.4 million barrels a day, would be in Saudi Arabia.[42]

All of these projections are subject to error. If they are assumed to be generally realistic, however, they suggest that Japan could not easily increase its imports of OPEC crude oil significantly without becoming more dependent on Middle Eastern oil, and probably also on Arab oil.

Japan might be able to increase its oil imports from non-Arab Iran somewhat by 1990. The advantage of doing so might not be great, however, since Iranian oil would be as vulnerable to new political or military disturbances as would Arab oil from the Middle East. Moreover, reducing dependence on Arab oil to a limited extent would not be too important. Japan sees that dependence less as a threat to its oil supplies than as a complication in its relations with the United States. Japan could placate the Arabs in a new Arab-Israeli crisis by adopting a more pro-Arab position, as it did in 1973. By doing so, however, it would widen the gap between its Middle Eastern policies and those of the United States.

Among the non-Arab members of OPEC outside the Middle East, only Indonesia and Ecuador are geographically well-located to export oil to Japan. Japan is not a natural market for Nigeria, Gabon, or Venezuela and has obtained little or no oil from these countries. In 1980 Japan imported 7,000 barrels of oil a day from Nigeria, 39,000 from Venezuela, and none from Gabon.[43]

42. Ibid., table 2.8, p. 16. These projections, from the midprice scenario, assume that Japan will import 6.2 million barrels of crude oil a day, slightly more than is projected in the central case of the present study.

43. U.S. Central Intelligence Agency (CIA), *International Energy Statistical Review*, December 22, 1981, p. 4. In June 1980 a joint venture of Mitsubishi Petroleum Development Company and France's ELF Aquitaine began production of 1,500 barrels a day from wells off the coast of Gabon. Production was expected to reach the still

Japan already receives about two-thirds of Indonesia's oil exports; almost all of the remainder goes to the United States.[44] Indonesia's domestic consumption is growing. If its productive capacity does not rise in the 1980s, Japan will be competing with the United States for declining Indonesian oil exports. Ecuador's productive capacity is expected to remain at only 0.2 million barrels a day.[45] In 1980 a Japanese company entered into a contract to buy 15,000 barrels a day of Ecuadorian oil.[46]

Japan may be able to increase its oil imports from non-OPEC suppliers during the 1980s, but probably not to an extent that would significantly reduce its dependence on Middle Eastern oil. Some OECD countries, such as Norway and the United Kingdom, will probably continue to be net oil exporters, but Japan is too far away to be one of their natural markets. The Alaskan oil that goes to the eastern part of the United States could more cheaply be shipped to Japan, but this is prevented by U.S. law. Countries that are not members of OPEC or the OECD (excluding Communist countries) have been projected to increase their oil production from 5.1 million barrels a day in 1979 to 10.5 million barrels a day in 1990. Over the same period, however, the oil consumption of these countries is projected to rise from 9.1 million barrels to 11.0 million barrels a day.[47] Some countries in this group would use their increased domestic production to reduce oil imports. Others would gain or add to an oil export capability, and it is from such countries that Japan might most easily obtain additional oil.[48]

The number of such countries is not great, and the amounts of oil potentially available do not appear to be large. The Japan National Oil Company (JNOC), a government corporation established to find and finance the development of oil resources in Japan and abroad, has concentrated its overseas activities in the Persian Gulf and Indonesia. JNOC has only four production projects in countries that are not

---

minuscule level of 7,000–8,000 barrels a day in 1981. Mitsubishi was reportedly considering swapping its share of the Gabon oil for Aquitaine oil from the Middle East. See *Japan Economic Journal*, July 22, 1980, p. 6.

44. CIA, *International Energy Statistical Review*, December 22, 1981, p. 4.

45. U.S. Department of Energy, *1980 Annual Report*, vol. 3, p. 10.

46. *Japan Economic Journal*, October 14, 1980, p. 6.

47. U.S. Department of Energy, *1980 Annual Report*, vol. 3, p. 16. Midprice scenario.

48. Importing small amounts of oil from countries that are net oil importers is not out of the question, but prospects obviously are best in countries with an export surplus.

members of OPEC: Egypt, Peru, Gabon, and Zaire. It also has exploration projects in Brazil, Canada, Malaysia, and Africa (country not specified).[49] The Japanese government has also tried, with limited success, to arrange for oil imports from Mexico.[50]

Oil exports to Japan from most of the countries mentioned above have thus far been too small to be listed separately in statistical compilations. The exceptions are Malaysia, which exported 77,000 barrels a day to Japan in 1981, and Mexico, which shipped to Japan 72,000 barrels a day in the same year.[51] Less complete information is available for the other countries.

Egyptian Petroleum Development Company, a Japanese firm, reportedly produced 5,000 barrels a day in 1980 from wells on the western coast of the Gulf of Suez and expected to double that production in 1981. This Japanese firm receives about 40 percent of output, or 2,000 barrels a day in 1980 and 4,000 barrels a day in 1981.[52] Japan reportedly began receiving 15,000 barrels a day from Peru in 1981, when the Japanese-financed trans-Andean pipeline was completed.[53] The limited results in Gabon have already been mentioned. The share of a Japanese company in the Zaire project (a joint venture with Gulf Oil) in 1980 was about 5,700 barrels a day. This oil was exchanged for Middle Eastern oil, which was shipped to Japan.[54]

Japan has long been interested in importing oil from the USSR and China. Imports of Soviet oil have thus far been negligible: only 12,000 barrels a day in 1982. Oil imports from China in 1982 were more substantial—179,000 barrels a day—but much less than was once hoped.[55] Prospects for significant oil exports to Japan from the USSR are poor. Total oil production in the Soviet Union is growing very slowly.[56] Rising

49. Japan National Oil Corporation, *JNOC* (Tokyo: JNOC, September 1981), pp. 8–11.

50. In the spring of 1980 Prime Minister Ohira visited Mexico and tried to obtain a commitment that oil exports to Japan would be increased to 300,000 barrels a day. Mexican President López Portillo refused to make this commitment unless Japan would make large investments in Mexican industrial development, investments that Ohira could not promise. See *Japan Economic Journal,* May 13, 1980, p. 2.

51. Japan External Trade Organization, *White Paper on International Trade: Japan, 1982* (Tokyo: JETRO, 1982), p. 188.

52. *Japan Economic Journal,* March 5, 1980, p. 7.

53. *Oil and Gas Journal,* vol. 79 (January 26, 1981), p. 98.

54. *JNOC,* p. 9.

55. CIA, *International Energy Statistical Review,* October 25, 1983, pp. 23, 27.

56. *Oil and Gas Journal,* vol. 79 (November 30, 1981), pp. 23–27, and vol. 80 (March 8, 1982), p. 102.

domestic requirements are reducing the margin available for export.[57] Moreover, most known Soviet oil reserves are located far from the Pacific ports through which oil exports to Japan would move.

Sakhalin Island, off the eastern coast of Siberia, is the one bright spot in this generally dark picture. A Soviet-Japanese venture has found oil and gas in two offshore areas and hopes to produce 150,000 barrels a day of oil beginning in 1988. This project, however, was delayed by the decision of the U.S. government, as part of its sanctions against the Soviet Union, to deny permission to export equipment needed to develop the oil and gas resources off Sakhalin.[58] The partial lifting of U.S. sanctions in November 1982 was expected to lead to the removal of this export ban.[59]

Oil production in China peaked at 2.118 million barrels a day in 1980 and then leveled off at about 2.0 million barrels a day in both 1981 and 1982. Exports (including small amounts of petroleum products) were 277,000 barrels a day in 1981 and 294,000 barrels a day in 1982.[60]

China is at a stage of economic development in which its requirements for petroleum products will almost certainly increase. There are no readily available and efficient substitutes for these liquid fuels in operating trucks, agricultural machinery, and small electric generating stations outside large power grids. If China is to expand its ability to export oil, domestic production must keep ahead of domestic requirements. This obviously calls for China to increase its efforts to find and to exploit domestic oil resources.

The most promising areas are in the far west of the country, in Hsinchiang (Xinjiang) Province, and offshore in the gulf of Pohai (Bohai), the Yellow Sea, the East China Sea, and the South China Sea. Because of Hsinchiang's great distance from major centers of population and industry, priority will probably continue to be given to offshore areas. Chinese oil exploration in some parts of these areas is complicated by conflicting legal claims with Japan and Korea in the north and with Vietnam in the south.[61]

---

57. *New York Times,* December 6, 1981.
58. *Japan Economic Journal,* February 23, 1982, p. 1.
59. Ibid., November 23, 1982, p. 1.
60. CIA, *International Energy Statistical Review,* March 29, 1983, p. 27.
61. The conflicting claims of Peking and Taipei are of a different order. The question is not one of China's rights, on which both might agree, but of who is entitled to speak for China.

Exploration in China's offshore areas appears to be increasing, with active Japanese participation. A Chinese-Japanese venture struck oil in the Gulf of Pohai in April 1981.[62] Japanese and other foreign firms have shown strong interest in obtaining drilling rights as the Chinese government opens more offshore areas for exploitation.[63] How much oil will be produced by this upsurge of activity cannot yet be determined. But a significant and steady increase in China's oil exports is a real possibility, and Japan should be able to obtain a substantial share of any increased oil exports from China.[64] Japan is the closest major market for Chinese oil, and the participation of Japanese firms in the development of China's oil resources will give them a claim on resultant oil production.

In summary, a review of possible sources of increased oil imports suggests that Japan has little chance of achieving more than a marginal decrease in its dependence on Middle Eastern oil by 1990. This dependence would actually increase unless about one-fourth, or about 400,000 barrels a day, of the increase in oil imports projected for the period 1980–90 is obtained outside the Middle East. For Japan to do so may not be easy.

Imports from OPEC members not in the Middle East could actually decrease—if, for example, Indonesia's ability to export is reduced by rising domestic consumption. The best prospects among non-Communist countries outside OPEC appear to be Malaysia and Mexico, but whether Japan could get an additional 400,000 barrels a day from these two sources is questionable. Significant oil imports from the USSR are not likely. The big question is China. If China could double its 1980 oil exports by 1990 to 600,000 barrels a day, Japan might be able to obtain half that amount, thereby increasing its imports from China by about 150,000 barrels a day. Increased imports from Malaysia, Mexico, and elsewhere might then be enough to prevent a rise in Japan's dependence on the Middle East.

The picture with respect to future dependence on Arab oil is similar. To avoid an increase in this dependence, about one-third, or over 500,000 barrels a day, of the increased oil imports projected for 1990 would have to come from non-Arab exporters. Mexico, Malaysia, and China would

62. *Oil and Gas Journal*, vol. 79 (May 25, 1981), p. 120.
63. Ibid., vol. 79 (November 9, 1981), p. 124; and *New York Times*, January 19, 1982.
64. Japan received 53.8 percent of China's oil exports in 1979 and 52.8 percent in 1980. See CIA, *International Energy Statistical Review*, December 22, 1981, p. 29.

Table 7-18. *Origins of Japan's Imports of Liquefied Natural Gas,*
Calendar Years 1978–81
Millions of metric tons

	1978		1979		1980		1981	
Country	Quantity	Percent	Quantity	Percent	Quantity	Percent	Quantity	Percent
Indonesia	3.7	33.0	6.2	44.9	8.5	50.6	8.7	51.2
Brunei	5.3	47.3	5.4	39.1	5.5	32.7	5.2	30.6
United Arab Emirates	1.3	10.7	1.2	8.7	1.9	11.3	2.0	11.8
United States	0.9	8.0	1.0	7.2	0.9	5.4	1.1	6.5
Total	11.2	100.0	13.8	100.0	16.8	100.0	17.0	100.0

Sources: Same as table 7-16. Numbers are rounded.

again be possibilities. In addition, oil imports from Iran might be increased. The result could be a modest, but not very significant, decrease in Japanese dependence on Arab oil.

Natural Gas

Japan's growing imports of liquefied natural gas have reduced its dependence on energy supplies from the unstable Middle East. In 1981 only 11.8 percent of imported liquefied natural gas was from the Middle East—and all of this was from Abu Dhabi, the leading member of the United Arab Emirates (table 7-18). Over half the imports of liquefied natural gas were from Indonesia, and three-tenths were from Brunei. The remaining 6.5 percent was from Alaska.

Earlier in this chapter Japan's imports of liquefied natural gas in 1990 were projected, under the assumptions of the central case, to be 46.2 billion cubic meters (approximately 35.5 million metric tons), over twice the amount of imports in 1981. Japan should have no difficulty in reaching this level of imports under existing and prospective contracts.

Japan has been importing liquefied natural gas under four contracts with these durations and annual targets (in millions of metric tons): Indonesia (1977–2000), 7.50; Brunei (1972–1992), 5.14; Abu Dhabi (1977–1997), 2.05; Alaska (1969–1984), 0.96.[65] The total—15.65 million metric tons—is, of course, valid only for those years in which all four projects are operating. Imports in 1981 were somewhat larger than the total amounts targeted (see table 7-18). The Alaskan contract will be extended to 1989.[66]

65. *Oil and Gas Journal*, vol. 79 (January 26, 1981), p. 99.
66. Conversation at the IEE (Tokyo, February 1982).

Five additional contracts for liquefied natural gas with these durations and annual targets (in millions of metric tons) are expected to be concluded: Malaysia (Sarawak) (1983–2003), 6.0; Indonesia (Bintang) (1983–2003), 3.2; Indonesia (Arun) (1984–2004), 3.3; Australia (1986–2005), 6.0; Canada (1986–2006), 3.0.[67] The annual targets of these contracts amount to 21.5 million metric tons.

The annual targets of all projects expected to be operating in 1990 (all existing and prospective projects except the one in Alaska) total 36.16 million metric tons, or about 0.7 million tons more than the total imports of liquefied natural gas projected for that year (see table 7-11). If the Alaskan project is extended again, the potential surplus would be on the order of 1.6 million metric tons a year. The estimate of imports of liquefied natural gas could be wrong, but these figures do suggest that Japan will not be under any strong compulsion to enter into contracts for liquefied natural gas during the 1980s in addition to those listed above.

Projects to develop supplies of liquefied natural gas that will probably be deferred include one in Qatar and two in the USSR (one on Sakhalin Island and one in the Yakutsk area of Siberia). Both the Qatar and Yakutsk projects are potentially very large, but plans for them are not as well advanced as the combined gas and oil project on Sakhalin, which could provide Japan with 3.5 million tons of liquefied natural gas annually beginning in 1985.[68] Japan has shown little interest in increasing its imports of liquefied natural gas from Alaska.[69]

If the geographic origins of Japan's imports of liquefied natural gas in 1990 are in proportion to the targets set in the contracts that are expected to be operative in that year, the structure (in percent) of Japan's imports of liquefied natural gas would be: Indonesia, 39.9; Malaysia, 16.3; Australia, 16.3; Brunei, 13.9; Canada, 5.4; Abu Dhabi, 8.2.

Uranium

Information on the international trade in natural uranium is difficult to obtain. The original (and continuing) reason for this secrecy was

67. Ibid.
68. *Japan Economic Journal*, November 11, 1980, p. 6.
69. In October 1982, the U.S. State Department reportedly asked the Japanese government whether Japan would be interested in importing liquefied natural gas from the Alaskan northern slope after 1990. The Japanese government reportedly declined to make an immediate reply because, if North Alaskan gas were imported, other commitments would have to be put aside. See Ibid., November 2, 1982, p. 1.

military. The controversy over the alleged existence of a uranium cartel in the 1970s provided an additional reason for not publicizing uranium shipments.

The lack of specific quantitative information on the amount and origins of Japan's uranium imports is therefore not surprising. It is generally known, however, that Canada is the largest supplier of uranium to Japan and that South Africa and Australia are also important suppliers. The bimonthly publication of the Japan Federation of Economic Organizations gave the following percentage breakdown of Japan's 1980 uranium imports by country of origin but gave no indication of physical magnitudes: Canada, 31.6; England, 22.4; South Africa, 14.8; Australia, 10.7; other, 20.5.[70] Because the United Kingdom is not a producer of uranium, the share attributed to England must represent the resale by British companies of uranium procured in other countries. The share attributed to South Africa may include uranium from Namibia.

Canada, South Africa (including Namibia), and Australia will probably continue to be Japan's principal suppliers of uranium throughout the 1980s and beyond. The only other important producers are the United States, Niger, and France. All or most of U.S. and French production will probably be used domestically. Australia's share in the Japanese market could increase because of the rapid increase projected in its attainable production capabilities.[71]

Before 1980, the U.S. Department of Energy was the sole supplier of uranium enrichment services to Japan. During the period 1980–89, Japan has contracted for 1.0 million separative work units (SWUs) a year, or a total of 10.0 million SWUs from the French-controlled Eurodif (Société Européenne d'Usine de Diffusion Gazeuse). (A separative work unit is a measurement of the work required to separate uranium isotopes in the enrichment process.) Over the same period, the Department of Energy estimates that it will provide 32.0 million SWUs to Japan, about three-fourths of Japan's requirements for enrichment services. Eurodif would supply the other one-fourth. The U.S. share would, however, rise from 65.5 percent in 1980 to 80.4 percent in 1989.[72]

Beginning in 1990, Japan's arrangements for imported enrichment

70. *Keidanren Review*, no. 72 (December 1981), p. 11.

71. See the estimates of attainable uranium production through 1990 in OECD Nuclear Energy Agency and International Atomic Energy Agency, *Uranium: Resources, Production and Demand* (Paris: OECD, December 1979), p. 23.

72. Personal communication from the U.S. Department of Energy (March 1982).

services become less certain. Japan may then have its own enrichment plant with an annual capacity of 1.0 million to 2.0 million SWUs. Whether Japan will enter into another contract with Eurodif or with some other enricher cannot be foreseen. Possibilities include the British-Dutch-German enterprise, Urenco, the USSR, and a possible new enrichment plant in Australia. In any event, the U.S. Department of Energy expects to remain a major supplier of enrichment services to Japan. It estimates average annual sales to Japan in the period 1990–95 of 4.6 million SWUs (compared with 4.1 million SWUs in 1989) and of 5.7 million SWUs in the period 1995–2000.[73]

Possible Developments in the 1990s

In the preceding chapter Japan's total final energy consumption in the year 2000 was projected by major consuming sectors under assumptions of the central case (table 6-17). These sectoral projections are only as valid as the assumptions on which they are based. GDP may grow more or less rapidly than the assumed 5 percent a year. Real energy prices may rise or fall rather than remain stable. The reactions of energy users to changes in income or price may be affected by unpredictable social and technological developments.

Even with these reservations, the central case projection of final energy consumption does provide a useful starting point for thinking about the future. As noted in chapter 6, this projection and the similar one for 1990 suggest that final energy consumption will increase more slowly during the last decade of this century than it was projected to increase during the 1980s. This conclusion is plausible because of the structural changes that appear to be taking place in the Japanese economy—the shift toward less energy-intensive activities.

A slower rate of growth in final energy consumption does not ensure a comparable slowdown in requirements for primary energy. Those requirements depend not only on final consumption, but also on transformation losses that are determined by the shares of different fuels in final consumption. The largest losses by far are incurred in generating electricity, so the proportion of energy consumed as electricity is especially critical.

73. Ibid.

Attempting a quantitative projection of the size and composition of primary energy requirements at the end of the century would be a highly speculative endeavor. The best that can be done is to try to determine the most likely direction, but not the exact magnitude, of change.

The first question to be considered is how the shares of different fuels in final energy consumption may change during the 1990s. Even if the shares of different fuels in the consumption of each sector were to remain the same, some change in the composition of total consumption would be brought about by the different rates of sectoral growth projected for the period 1990–2000. The projected decline in the relative importance of the industrial sector, and the projected increase in the relative importance of the transportation sector, would cause a small but significant (about 1 percent) shift from coal to oil.

The shares of different fuels in the consumption of all sectors would almost certainly change. The transportation sector would probably continue to rely overwhelmingly on oil, and the small nonenergy sector is by definition confined to certain petroleum products. Some decline in the oil dependence of the industrial and residential and commercial sectors, however, appears likely. In both sectors the fuel substituted for oil would probably be electricity.

In the industrial sector, the shift from oil to electricity would reflect both the move away from basic materials to higher technology products and the decline in the domestic production of naphtha-based intermediate petrochemicals. In the residential and commercial sector, the importance of electricity could increase because of greater use of air conditioning and, possibly, of heat pumps to heat and cool space.

The second question to be considered is how the electricity entering into final energy consumption will be generated. If—as suggested above—the share of electricity in final energy consumption increases, this question will become more important in the 1990s than it is in the 1980s.

Earlier in this chapter primary electric power plants—nuclear, hydroelectric, and geothermal—were projected to produce 28.8 million t.o.e. of electricity in 1990, or 38.5 percent of total estimated electricity generation in that year. To keep pace with the projected 30 percent growth in total final energy consumption from 1990 to 2000, primary electricity production would have to increase to 37.4 million t.o.e. Most of this increase would have to come from expanded nuclear capacity.

Hydroelectric capacity might rise slightly to 25 gigawatts from the 23 gigawatts projected for 1990, and geothermal capacity could double (1 gigawatt is projected for 1990). Hydroelectric and geothermal plants could then produce 12.2 million t.o.e. of electricity. If nuclear plants produced the remaining 25.2 million t.o.e., nuclear generating capacity would have to be about 50 GW(e), substantially more than the capacity of 36 GW(e) projected for 1990.[74] Attaining this nuclear capacity by the year 2000 is quite feasible but by no means assured.[75] Rising construction costs and public opposition could force Japanese utilities to turn to conventional thermal plants for added capacity.

If—as appears likely—the share of electricity in Japan's final energy consumption increases during the 1990s, primary electricity would be a declining percentage of total electricity unless it did more than merely keep pace with the growth of total final energy consumption. In any event, it appears certain that conventional thermal plants will continue to supply a large part of total electricity requirements. The third question to be considered is, therefore, what fuels will be used in conventional thermal plants in the 1990s.

A continued decline in Japan's reliance on oil to generate electricity is likely, but how steep the decline will be is not certain. A slowing in the rate of substitution of coal and gas for oil is entirely possible. The shift away from oil is propelled by government policy, reinforced by the high price of oil. A period of stability in the Middle East could cause the Japanese government to place less emphasis on reducing dependence on oil from that part of the world. Price would then become the main influence on the utility companies' choice among hydrocarbon fuels. If, as is possible, real oil prices remain stable or decline, and if, as is likely, real coal prices rise, the interest of the utilities in coal could weaken.[76] The price advantage of liquefied natural gas over oil is narrow at best. If oil became clearly cheaper, more generating capacity fueled by liquefied natural gas would be attractive only as a means of meeting environmental standards in heavily populated areas.

74. One gigawatt of hydroelectric or geothermal capacity is assumed to produce 450 thousand t.o.e. of electricity a year; 1 gigawatt of nuclear capacity is assumed to produce 500 thousand t.o.e. a year.

75. There appear to be no authoritative, recent estimates of nuclear generating capacity in the year 2000. The International Nuclear Fuel Cycle Evaluation (INFCE) estimated that Japan would have 100–150 GW(e) of nuclear capacity in 2000. See OECD Nuclear Energy Agency and International Atomic Energy Agency, *Uranium*, 1979 ed., p. 29. Even the lower end of the INFCE range appears to be much too high.

76. Japanese utilities have never been enthusiastic about shifting to coal. Obtaining assured supplies of coal often requires investing in foreign mines. Coal is also more difficult to handle than oil and creates more serious air pollution problems.

A final question is what may happen to Japan's requirements for imported energy in the 1990s.

Requirements for natural uranium and uranium enrichment services will rise with nuclear generating capacity. The need for imported uranium may be somewhat reduced, however, by recycling plutonium and residual uranium extracted from spent fuel. Recycling plutonium would also reduce the need for enrichment services.[77] A fraction of requirements for enrichment services will probably be provided during the 1990s by a new domestic facility.

Domestic production of hydrocarbons will probably remain low. Requirements for imported hydrocarbons will thus depend mostly on the amounts of these fuels needed for direct consumption and for electricity generation. As a result of the shift from oil to electricity, direct consumption of hydrocarbons could grow less rapidly than total final energy consumption. This shift would, however, almost certainly cause a net increase in hydrocarbon consumption because of the large transformation losses in generating electricity.[78] This effect would be reduced, however, to the extent that electricity is generated in primary electric plants.

If requirements for imported hydrocarbons grow 30 percent from 1990 to 2000 (the same rate of growth projected for total final energy consumption), requirements in the year 2000 would be about 560 million t.o.e. How this hypothetical total would be divided among coal, oil, and gas cannot be foreseen. If the share of oil were 60 percent (compared with the 71 percent projected for 1990), oil import requirements in 2000 would be about 336 million metric tons, or roughly 10 percent more than projected requirements in 1990. This modest increase of approximately 600,000 barrels a day might be obtained from China, Mexico, and other suppliers outside the Persian Gulf.

If 27 percent of hydrocarbon import requirements were assigned to coal and 13 percent to liquefied natural gas (requirements of these fuels were projected to be 19 percent and 10 percent, respectively, in 1990), coal requirements in 2000 would be 151 million t.o.e., and liquefied

---

77. Japanese utilities have recently shown a strong interest in recycling plutonium in their present generation of light water reactors. It had previously been assumed that plutonium extracted from spent fuel would be held for later use in the advanced thermal and fast breeder reactors that are still being developed.

78. At 37 percent generating efficiency, every kilocalorie of electricity produced requires an input of about 2.7 kilocalories.

natural gas requirements would be 73 million t.o.e. The increase above
the 1990 projections of the central case would be 86 percent for coal and
63 percent for liquefied natural gas. In original physical units, import
requirements for coal would be 220 million metric tons, an increase of
102 million tons. Import requirements for liquefied natural gas would be
74.5 billion cubic meters, an increase of 28.7 billion cubic meters.[79]

Japan should be able to procure these increases in imported coal and
liquefied natural gas. Countries that export little or no coal today might
supplement imports from established suppliers. Colombia, Indonesia,
and African countries such as Chad, Niger, Zaire, Namibia, Angola, and
Bostwana are possible sources. Additional liquefied natural gas might
be obtained both from existing suppliers and from new projects that were
considered but not undertaken in the 1980s, such as the projects in Qatar
and Siberia (Sakhalin and Yakutsk).

The calculations for the future that have been made in this chapter
are purely illustrative. Reality could turn out quite differently. Critical
uncertainties include the rate of growth of Japan's final energy consumption,
the share of electricity in final consumption, and the percentage of
electricity that will be produced in primary electric plants rather than in
conventional thermal plants.

79. Liquefied natural gas was converted at the rate of 9,800 kilocalories per cubic
meter. Coal was converted at the rate of 6,850 kilocalories per kilogram (the average
of the rates for steam and coking coal).

CHAPTER EIGHT

Future Energy Developments in Taiwan

THE TRENDS in energy supply and use in Taiwan in the years following the 1973–74 oil crisis will not continue throughout the 1980s. Increasing supplies of nuclear energy and imported coal will reduce the past heavy dependence on oil. The rise in the overall energy intensity of the economy also appears to have ended and will probably be followed by a gradual decline.

The official four-year economic plan provides a useful basis for considering possible energy developments in the first half of the 1980s.[1] No comparable guide exists for the remainder of the decade, although the now outdated ten-year plan contains some useful insights.[2]

Future Energy Requirements

Table 8-1 shows a rise in the overall energy intensity of Taiwan's economy after the first oil crisis, a leveling off of energy intensity in 1979–80, and a sharp drop in energy intensity in 1981. During the period 1973–79 the depressing effect of higher real energy prices on energy consumption was overcome by the increased energy requirements created by several major construction projects and by the expansion of

1. Council for Economic Planning and Development, *Four-Year Economic Plan, 1982–85* (Taipei: CEPD, December 1981; in Chinese). Hereafter CEPD, *Four-Year Plan*.
2. CEPD, *Ten-Year Economic Development Plan for Taiwan, Republic of China (1980–89)* (Taipei: CEPD, March 1980).

Table 8-1. *Overall Energy Intensity of the Economy of Taiwan
in Selected Years, 1973–81*

Year	Gross domestic product (billions of 1976 New Taiwan dollars)	Primary energy supply (ten billions of kilocalories)[a]	Energy intensity (kilocalories per 1976 New Taiwan dollar)
1973	582.1	14,205	244.0
1976	701.1	18,974	270.6
1979	940.6	26,763	284.5
1980	1,004.6	28,607	284.7
1981	1,060.0	27,310	257.6

Sources: GDP: Directorate-general of Budget, Accounting, and Statistics, *National Conditions* (Taipei: Statistical Bureau, Winter 1981), p. 32. Primary energy supply: figures for 1973–80 are the totals shown for internal final consumption in Energy Committee, Ministry of Economic Affairs, *Taiwan Energy Statistics, 1980* (Taipei: The Ministry, 1981), pp. 133, 139, 145, 147; figures for primary energy supply 1981 are from Council for Economic Planning and Development (CEPD), *Four-Year Economic Plan, 1981–85* (Taipei: CEPD, December 1981; in Chinese), table 8-1, p. 122. Numbers are rounded.

a. Ten billion kilocalories approximately equal a thousand metric tons of oil equivalent (t.o.e.).

energy-intensive industries, including petrochemicals and steel. In 1979–80 a new upsurge in energy prices began to tip the balance the other way. The drop in energy intensity in 1981 was in part attributable to short-run and probably transitory developments: a slowing of economic activity and an associated, absolute reduction in total energy use. Some decline in energy intensity in 1981 was to be expected, however, even if the economy had grown more rapidly. In time, higher energy prices and the pressure of competition from foreign producers with lower energy costs were bound to bring about structural changes that would reduce the overall energy intensity of Taiwan's economy.

The Four-Year Plan

The four-year plan for the period 1982–85 is a recent analysis of economic trends in Taiwan by well-qualified analysts. The plan's estimates of energy use in 1985 will be adopted as the central case projections in this study. These estimates are of primary energy requirements. Transformation losses in making coal and petroleum products are shown separately but are not distributed among the consuming sectors. The much greater transformation losses in generating electricity are not shown separately and are included in the plan's sectoral totals.[3]

3. These attributes of the four-year plan's projections make it difficult to estimate final energy consumption by sector, as is done in the chapters dealing with possible future developments in Japan (chapter 6) and Korea (chapter 9).

Table 8-2. *Primary Energy Requirements in Taiwan under the Four-Year Economic Plan, 1981–85*
Ten billions of kilocalories[a]

	1981		1985		Annual rate of change (percent)
Sector	Quantity	Percent	Quantity	Percent	
Agriculture	995	3.6	1,167	3.2	4.07
Industry	14,252	52.2	18,321	50.2	6.48
Residential and commercial	5,094	18.7	6,681	18.3	7.02
Transportation	3,063	11.2	4,312	11.8	8.93
Nonenergy uses of fuels[b]	2,624	9.6	4,113	11.3	11.89
Energy loss[c]	1,282	4.7	1,909	5.2	10.47
Total	27,310	100.0	36,503	100.0	7.52

Source: CEPD, *Four-Year Plan*, table 8-1, p. 122. Original figures in kiloliters of oil equivalent have been converted at the rate of 9,000 kilocalories per liter. Exports and stock changes shown in the sources have not been included here, since they are not part of primary energy requirements. Numbers are rounded.

a. Ten billion kilocalories approximately equal a thousand t.o.e.

b. Nonenergy requirements include petroleum products and natural gas used as chemical feedstocks.

c. "Energy loss" is assumed to be transformation losses in making petroleum and coal products. Transformation losses in generating electricity are included in sectoral figures.

The four-year plan projects a continued modest decline in energy intensity. Total primary energy requirements are projected to increase at an average annual rate of 7.52 percent (table 8-2). The gross domestic product is assumed to grow at 8.0 percent annually. Overall energy intensity would therefore decline from 257.6 kilocalories per 1976 New Taiwan dollar in 1981 (see table 8-1) to 252.9 kilocalories per 1976 New Taiwan dollar in 1985.[4] The projected decline in energy intensity from 1981 to 1985 would be only 1.75 percent. The decline from 1980 to 1985, however, would be 11.13 percent, which is a more meaningful comparison because 1981 was an abnormal year. The 7.52 percent annual rate of increase in total energy use projected for the period 1981–85 is below the assumed 8.0 percent annual growth of GDP. From 1973 to 1979, however, energy use rose 11.3 percent annually, whereas GDP was growing 8.33 percent a year.

On the one hand, the 8.0 percent annual average growth of GDP assumed in the plan may prove to be high if the recovery of Taiwan's export markets from the recession of the early 1980s is sluggish.[5] On the

4. Under the plan, total primary energy requirements in 1985 would be 36.5 million metric tons of oil equivalent (t.o.e.) and GDP would be 1,442.1 billion 1976 New Taiwan dollars.

5. During 1982, the first year of the plan, the economy of Taiwan grew 3.8 percent. CEPD estimates growth in 1983 at 7.1 percent and forecasts 7.5 percent growth in 1984. See *Asian Wall Street Journal Weekly*, May 9, 1983, p. 2, and December 26, 1983, p. 7.

Table 8-3. *Changes in the Structures of GDP and Energy Use in Taiwan under the Four-Year Economic Plan, 1981–85*
Percent

	1981		1985	
Sector	GDP	Energy use[a]	GDP	Energy use[a]
Agriculture	7.3	3.8	5.9	3.4
Industry	50.6[b]	64.8[c]	51.5[b]	64.9[c]
Transportation[d]	6.5	11.8	7.0	12.4
Other	35.6	19.5	35.6	19.3
Total	100.0	100.0	100.0	100.0

Source: CEPD, *Four-Year Plan*, table 2-2, p. 11, and table 8-1, p. 122. Numbers are rounded.
  a. The economic coverage of sectors in energy use is presumably the same as in GDP. The source has included energy used in generating electricity in the appropriate consuming sectors at its input value. Energy lost in making coal products and petroleum products (see table 8-2) has been arbitrarily allocated among consuming sectors in proportion to sectoral shares of total energy use.
  b. GDP figures for industry include mining, manufacturing, construction, and electricity and other utilities.
  c. Nonenergy uses of fuels have been included in energy use by industry.
  d. Transportation includes communications.

other hand, the plan's assumption of a 3.0 percent average annual increase in international oil prices is almost certainly too high, and this would help to validate the assumption about growth of GDP.

The four-year plan projects quite different growth rates for the various energy-consuming sectors (table 8-2.) The highest rate of increase (in nonenergy uses of fuels) is nearly three times that of the lowest (in agriculture). Energy use in transportation is also expected to grow rapidly. The rates of growth in energy use projected for the industrial sector and for the residential and commercial sector are, however, below the rate of growth for the economy as a whole.

Because the plan covers only four years, the changes in the structure of energy use produced by the differences in sectoral rates of growth are not dramatic. As would be expected, the shares foreseen for nonenergy uses and transportation in 1985 are larger than in 1981, and the shares of the other three sectors are smaller (table 8-2).

Table 8-3 compares the changes in the structures of economic production and energy use set forth in the four-year plan. It is difficult to match data for economic output and for energy use except in broad categories, so some significant structural changes are undoubtedly missed in this kind of comparison. The share of the "other" category in both GDP and energy use is expected to remain virtually the same over the period of the plan. The share of agriculture in both GDP and energy use is expected to decline, and increases are foreseen in the share of transportation in both categories.

Table 8-4. *Future Energy Consumption in Taiwan Estimated under Ten-Year and Four-Year Plans, 1985*
Ten billions of kilocalories[a]

Sector	A. Ten-year plan	B. Four-year plan	B/A (percent)
Agriculture	1,490	1,167	78.3
Industry	25,589	18,321	71.6
Transportation	4,280	4,312	100.7
Residential and commercial	10,458	6,681	63.9
Nonenergy uses of fuels	6,822	4,113	60.3
Energy loss	n.a.	1,909	n.a.
Total	48,639	36,503	75.0

Sources: CEPD, *Ten-Year Economic Development Plan* (Taipei: CEPD, March 1980), p. 58, and *Four-Year Plan*, table 8-1, p. 122. Figures in thousands of kiloliters of oil equivalent have been converted at the rate of 9,000 kilocalories per liter. The ten-year plan gives projections for 1984 and 1986. Estimates for 1985 were obtained by interpolation. The ten-year plan does not show a separate figure for energy lost in making coal products and petroleum products. Numbers are rounded.
n.a. Not available.
a. Ten billion kilocalories approximately equal a thousand t.o.e.

For agriculture and transportation, changes in economic importance partially explain changes in shares of energy use. This explanation definitely does not hold true for the industrial sector. The share of industry in GDP is projected to increase from 50.6 percent in 1981 to 51.5 percent in 1985. Industry's share of total energy use (including nonenergy use of fuels), however, is expected to remain almost the same (64.8 percent in 1981 and 64.9 percent in 1985).

The expectations of economic planners in Taipei concerning future energy consumption changed substantially during the period of less than three years between publication of the ten-year plan and publication of the four-year plan. Table 8-4 gives the comparison. Part of the difference between the two sets of estimates can be explained by the failure of the economy to grow as rapidly in 1980 and 1981 as had been assumed in the ten-year plan. GDP in 1981, the base year for the four-year plan, was 9.7 percent lower than had been projected in the ten-year plan.[6] The ten-year plan also could not take full account of the effects of the increases in energy prices in 1979 and 1980. By the time the four-year plan was

6. Both plans assumed that Taiwan's economy would grow at an average annual rate of about 8.0 percent. The ten-year plan assumed 8.0 percent annual growth from 1980 to 1984 and 7.8 percent annual growth from 1985 to 1989. The four-year plan assumed 8.0 percent annual growth for the entire planning period, 1981–85. The actual rates of growth of GDP in 1980 and 1981 were 6.8 percent and 5.5 percent, respectively.

prepared, the full extent of those price increases was known, and the government had had time to make appropriate adjustments in its policies.

The 25 percent reduction in the projection of total energy use in 1985 can be seen as simply an estimate of the reactions of consumers to higher energy prices. The reduction was, however, distributed unevenly among consuming sectors, and this uneven distribution reflects a changed view of the future structure of production. The two plans predicted the following percentage sturctures of GDP in 1985:[7]

	Ten-year plan	Four-year plan
Agriculture	6.7	5.9
Industry	56.1	51.5
Transportation	6.1	7.0
Other	31.1	35.6

The scaled-down economic importance of industry is consistent with the large percentage reductions in the estimates of energy consumption in the industry and nonenergy sectors.

The changed view of the future structure of Taiwan's economy is also quite consistent with the goals of the government. At a seminar on Taiwan's foreign trade held in August 1981, Yu Kuo-hwa, chairman of the Council for Economic Planning and Development, called for a more rapid change in industrial structure to sustain the competitiveness of Taiwan's exports and deal with serious energy problems. Yu said that "priority should be given to the development of such strategic industries as electronics machinery, information, precision equipment, and automobile production. These industries consume less energy and they are value-added industries." Yu also called for an end to the expansion of the petrochemical industry, except for specific domestic needs. He said that exporters of man-made fiber and plastic products should be allowed to buy cheaper intermediate petrochemicals abroad so that they might maintain their international competitiveness.[8]

Energy Requirements after 1985

The four-year plan is not intended to establish long-term trends but to facilitate a change in economic structure. The plan's projections

7. CEPD, *Ten-Year Economic Development Plan*, gives the sectoral distribution of GDP in 1984 and 1989 in table 3, p. 13. The distribution in 1985 was obtained by interpolation. The figures for the four-year plan are from table 8-3 of this chapter.

8. *China News*, August 22, 1981, p. 4.

cannot, therefore, automatically be extended beyond the 1981–85 planning period. For one thing, a somewhat lower rate of economic growth—say, 7.0 percent a year—may be a reasonable assumption for the second half of the 1980s.[9] The plan's implicit assumption of fairly stable energy prices can, however, be continued.[10]

Given an assumption concerning the rate of growth of GDP, the most important question is what will happen to the overall energy intensity of the economy as measured by the ratio of energy use to GDP. The four-year plan, as noted, projected an 11.13 percent decrease in this ratio from 1980 to 1985, or an average annual rate of decline of 2.16 percent.[11] Whether energy intensity will change at a different rate in the period 1985–90 depends both on general factors and on factors affecting particular consuming sectors. Among the general factors are energy prices, thermal efficiency, the net energy content of nonfuel trade, and energy transformation losses.

An inverse relationship exists between real energy prices and energy efficiency. The stable energy prices assumed for the period 1985–90 (instead of the small increase posited in the four-year plan) would therefore slightly retard the decline in energy intensity.

The shift from oil to coal would tend to reduce thermal efficiency somewhat. Modern coal-burning equipment, however, has narrowed the difference between the thermal efficiencies of these two fuels. The four-year plan projects a substantial substitution of coal for oil from 1980 to 1985.[12] The probable continued substitution of coal for oil would have to proceed at a significantly different rate from 1985 to 1990 to have much effect on the rate of change in overall energy intensity.

The net energy content of nonfuel trade depends on the difference in the values of nonfuel exports and nonfuel imports and on the energy

9. Conversations with economic analysts in Taipei, April 1982.

10. The four-year plan's assumption of 3.0 percent annual increase in international oil prices would bring about a much smaller rate of increase in domestic prices of refined petroleum products and still smaller increases in the domestic prices of electricity and other fuels.

11. Because 1981—the actual base year for the four-year plan—was an abnormal year, 1980 is used here in calculating the change in energy intensity foreseen in the plan.

12. CEPD, *Four-Year Plan* (table 8-1, p. 122), has the share of coal rising from 14.7 percent of total energy supply in 1981 to 25.3 percent in 1985 and the share of oil falling from 66.8 percent to 52.6 percent over the same period. A shift of this magnitude in only four years may, of course, not be feasible.

intensities (in kilocalories per New Taiwan dollar) of those exports and imports. The four-year plan projects a narrowing of Taiwan's relatively small favorable balance of trade in all goods and services (including fuels) from 1981 to 1985.[13] The plan does not, however, indicate whether the favorable balance of nonfuel trade is expected to increase or decrease, and, if so, by how much.

If the overall trade balance is in equilibrium (as it is projected almost certain to be in 1985), the excess of nonfuel exports over nonfuel imports must equal the fuel import bill. The continued rise in requirements for imported energy projected in the four-year plan and the assumed 3.0 percent annual increase in international oil prices would far outweigh the savings achieved by substituting cheaper fuels (in heat content per dollar) for some imported oil. The fuel import bill therefore must rise over the period of the four-year plan, and the excess of nonfuel exports over nonfuel imports must increase. The net exports of embodied energy largely depend on the size of this excess, but the excess would also be influenced by the relative energy intensities of nonfuel exports and nonfuel imports.

It is not clear whether the four-year plan foresees net exports of embodied energy growing more or less rapidly than total energy requirements. In the former case they would retard the decline in overall energy intensity; in the latter case they would accelerate it. In the period 1985–90 the rate of growth of requirements for imported fuels will again be the dominant factor. The assumed stability of international oil prices would, however, moderate the increase in the fuel oil bill, and a more rapid displacement of oil by cheaper fuels would work in the same direction. The energy embodied in net nonfuel exports may also be somewhat reduced as Taiwan exports more high-value-added, low-energy-intensive goods and increases its imports of high-energy-intensive commodities such as aluminum ingots and intermediate petrochemical products. The combined effect of these changes should be to favor a further decline in the energy intensity of the economy.

Transformation losses in making coal products and petroleum products are projected in the four-year plan to grow more rapidly than total energy requirements. The share of these transformation losses in total requirements, however, is projected to increase from only 4.3 percent

13. Ibid., table 2-1, p. 10. The favorable balance is 1.19 percent of total trade in 1981 and 0.62 percent in 1985.

in 1981 to 5.0 percent in 1985.[14] Their effect on overall energy intensity would therefore not be great. In the second half of the 1980s these losses should level off, and may decline, as the rate of increase in the use of petroleum products slows down and as completion of planned steel capacity at least temporarily checks the rise in coke requirements.[15]

Transformation losses in generating electricity are an important determinant both of total energy requirements and of overall energy intensity. In 1981 the Taiwan Power Company used about 2,400 kilocalories of energy in its thermal power plants to produce 1 kilowatt of electricity.[16] If, as is customary, 1 kilowatt of electricity is given the energy value of 860 kilocalories, generating efficiency was 35.8 percent.[17] That is, each unit of energy in the form of electricity required the use of 2.79 units of some other form of energy. Transformation losses were 1.79 units of energy per unit of electrical energy.

Taiwan energy statistics give electricity the value of the energy used to produce it. In 1980 electricity had a total energy value of 10.1 million t.o.e.[18] The four-year plan estimates that in 1985 electricity will have a total energy value of 12.9 million t.o.e. At 35.8 percent generating efficiency, transformation losses in 1980 would be 6.5 million t.o.e., or 20.4 percent of total energy requirements. In 1985 they would be 8.3 million t.o.e., or 22.7 percent of total requirements. The annual rate of increase of transformation losses in generating electricity from 1980 to 1985 would be 5.07 percent.

The increase in the share of these transformation losses in total energy requirements caused the decline in energy intensity projected in the four-year plan to be somewhat more gradual than would otherwise have been the case. The question for the second half of the 1980s is whether the share of electrical transformation losses in total energy requirements

14. Ibid., table 8-1, p. 122.

15. The initial annual capacity of China Steel Company was 1.5 million metric tons. Completion of a second blast furnace in 1982 raised capacity to 3.3 million tons. Further increases in capacity to 5.65 million tons in 1985 and 8.0 million tons in 1988 are planned. See *Asian Wall Street Journal Weekly*, May 18, 1981, p. 17, and *China News*, February 20, 1982, p. 4.

16. Taiwan Power Company, *Taipower Annual Report, 1981* (Taipei: Taipower, 1981), p. 15.

17. Taiwan Power Company, *Annual Report, 1980*, p. 14. Taiwan Power Company reported transmission and distribution losses of 6.83 percent in 1980. In the interest of simplicity, these losses will be ignored in the present analysis.

18. Energy Committee, Ministry of Economic Affairs, *Taiwan Energy Statistics, 1980* (Taipei: The Ministry, 1981), p. 147.

will rise more or less rapidly than they are projected to rise in the first half of the decade.

Because generating efficiency appears to be near its technological limit, the rate of increase in electrical transformation losses will for the most part depend on the rate of increase of electricity use. Electricity consumption is projected in the four-year plan to increase at an average annual rate of 7.6 percent. Consumption in 1981, however, was 1.3 percent below consumption in 1980.[19] If 1980 is taken as the base year, the projected annual rate of increase is 5.7 percent. A somewhat higher rate of increase in the second half of the 1990s would not be surprising as industry emphasizes higher-technology products and as air conditioning spreads to more households and commercial establishments. The share of electrical transformation losses in total energy requirements may therefore increase somewhat.

It is not possible to express quantitatively the net influence on overall energy intensity of the general factors reviewed above. Reductions in Taiwan's net exports of embodied energy and in nonelectrical transformation losses favor a somewhat greater rate of decline in energy intensity. Stable energy prices and increased electrical transformation losses work in the opposite direction. The influence of a possible change in thermal efficiency is uncertain, but it probably would not be large.

An alternative to the above approach is to consider whether the rates of change in sectoral ratios of energy use to GDP implicitly projected in the four-year plan are likely to continue into the late 1980s. Actual ratios for 1980, and ratios for 1985 derived from the four-year plan, are set forth in table 8-5.

The figures in the table illustrate one remarkable circumstance: almost all of the decrease in the energy intensity of Taiwan's economy foreseen in the four-year plan can be traced to the industrial sector. The effects on overall energy intensity of projected declines in the ratios of energy use to GDP of the agricultural, residential and commercial, and non-energy sectors are quite minor. The ratio for the transportation sector is projected actually to rise slightly.

The 4.0 percent annual decline in the ratio of energy use to GDP of the industrial sector projected for the first half of the 1980s contrasts with the increase in the same ratio during the 1970s. From 1973 to 1976 the ratio increased at an annual rate of 1.4 percent. From 1976 to 1979

19. Taiwan Power Company, *Annual Report, 1982*, p. 34.

Table 8-5. *Estimated Annual Rates of Change in Sectoral Ratios of Energy Use to GDP in Taiwan, 1980–85 and 1985–90*

	Ratio of energy use to GDP (kilocalories per 1976 New Taiwan dollar)					Annual rate of change (percent)	
Sector	1980	1985	Differ-ence (1980–85)	1990	Differ-ence (1985–90)	1980–85	1985–90
Agriculture	9.4	8.5	−0.9	7.7	−0.8	−2.0	−2.0
Industry	164.7	134.0	−30.7	115.7	−18.3	−4.0	−2.9
Residential and commercial	50.2	48.8	−1.4	47.4	−1.4	−0.6	−0.6
Transportation	30.1	31.5	1.4	32.9	1.4	0.9	0.9
Nonenergy uses of fuels	30.3	30.1	−0.2	28.2	−1.9	−0.1	−1.3
Total	284.8	252.9	−31.8	231.9	−21.0	−2.3	−1.7

Sources: Sectoral energy consumption figures used in calculating actual 1980 ratios are from Energy Committee, *Taiwan Energy Statistics, 1980*, p. 147. GDP in 1980 is from Directorate-general of Budget, Accounting, and Statistics, *National Conditions* (Taipei: Statistical Bureau, Winter 1981), p. 32. Sectoral energy consumption estimates for 1985 are from CEPD, *Four-Year Plan*, table 8-1, p. 122. GDP in 1985 was projected on the assumption of an 8.0 percent annual rate of growth from 1981 to 1985. Energy losses in making coal products and petroleum products are not shown separately for 1980. The ratio of such losses to GDP projected for 1985 in the four-year plan is 13.2 kilocalories per 1976 New Taiwan dollar; this figure has been distributed proportionately among all consuming sectors. Numbers are rounded.

the annual rate of increase speeded up to 1.8 percent. The ratio then leveled off and was only slightly higher in 1980 than it was in 1979.[20]

Even if the sharp drop in the ratio in 1981 is ignored,[21] something new appears to have been happening in the industrial sector at the end of the 1970s and the beginning of the 1980s. The authors of the four-year plan in effect concluded that this new factor was the onset of a decline in the amount of energy that must be consumed in manufacturing to produce a unit of GDP. This conclusion rests on the belief—itself a major feature of the plan—that the government should encourage important structural changes in Taiwan's manufacturing industries and that such changes are indeed under way.

The four-year plan divides manufacturing into "key manufacturing" and "other manufacturing." The share of all manufacturing in GDP is projected to increase only from 39.7 percent in 1981 to 40.3 percent in 1985. The share of key manufacturing in GDP, however, is projected to

20. Energy use figures used in calculating these ratios are from Energy Committee, *Taiwan Energy Statistics, 1980*, pp. 132–45. GDP figures are from Directorate-general of Budget, Accounting, and Statistics, *National Conditions* (Taipei: Statistical Bureau, Winter 1981), p. 16.

21. The four-year plan figures for 1981 suggest that the industrial ratio of energy use to GDP in that year was lower than it was in 1973.

increase from 10.6 percent to 13.2 percent over the same period. In 1981 key manufacturing accounted for 26.6 percent of all manufacturing production. In 1985 it is projected to account for 32.9 percent.[22]

Key manufacturing industries are defined as including metal products, machinery, electrical machinery, electronics, electrical appliances, and means of transportation.[23] The plan also refers to "policy-guided industries," which appear to be the same as key manufacturing industries. The plan declares that—to accelerate the upgrading of industry, improve industrial structure, decrease the intensity of energy consumption in industry, and increase productivity—future development will be focused on those policy-guided industries such as machinery (including general machinery, electrical machinery, sophisticated and automotive machinery, and transportation machinery), and telecommunications (including computer software, microcomputers and associated equipment, digital telecommunications, and related electronics).[24]

The energy chapter of the four-year plan relates the new industrial strategy to the goal of reducing energy intensity. This chapter explains that the high energy intensity of Taiwan's economy was the result of past emphasis on petrochemicals, basic metals, cement, and glass. The new strategy is to shift to the development of industries with low energy intensity, such as electronics, computers, data processing, electrical machinery, and sophisticated chemicals. Transforming the structure of industry will, however, take time. The energy intensity of the economic system will gradually decrease with the control of energy-intensive industry and the development of technology-intensive industry.[25]

The four-year plan sets fairly specific goals for both strategic (or key) and other industries. The means of reaching those goals, however, are, not spelled out in any detail. In general, the strategic industries are to be promoted through loans, government investment, tax preferences, and government support for the development and importation of new technology. The plan devotes relatively little discussion to the "control" of energy-intensive industries.[26] The desired direction of development of energy-intensive industries is in some cases indicated rather obliquely. Thus domestic fertilizer supply is labeled "adequate," and cement is

22. CEPD, *Four-Year Plan*, table 2-2, p. 11.
23. Ibid., pp. 10–11.
24. Ibid., pp. 48–51.
25. Ibid., pp. 117–19.
26. Ibid., pp. 51–65.

described as "mainly for domestic use." The plan does, however, clearly call for importing cheaper substitutes for at least part of the domestic production of aluminum ingots and intermediate petrochemicals. No changes are proposed in the expansion plans of the steel industry, and the copper industry is assumed to expand to meet domestic needs.[27]

The structural changes in manufacturing industries contemplated in the four-year plan will clearly still be under way in the second half of the 1980s. Whether the rate of change will accelerate or slow down is an open question. Even if it remains fairly steady, the ratio of energy consumed in industry to GDP could increase or decrease. Taking the assumed annual rates of growth in GDP as given (8.0 percent from 1981 to 1985 and 7.0 percent for the remainder of the decade), the direction of change in the ratio depends on energy consumption.

An increase in the share of key industries in total manufacturing will tend to reduce the rate of increase of energy consumption. How this increase comes about makes a difference. If it is the result of a decline in the share of energy-intensive industries, the effect on energy consumption will probably be greater than if it is the result of an expansion of the output of key industries themselves. This is the case because the technology-intensive key industries rely more on electricity, with its high transformation losses, than do other manufacturing industries.

In the first half of the decade structural change in manufacturing may depend more on the decline of energy-intensive industries, such as aluminum, than on the expansion of key industries, such as electronics. In the second half of the decade the relative importance of these two instruments of structural change may be reversed. The market forces depressing energy-intensive industries are already present, but the expansion of technology-intensive industries takes time. Everything else remaining equal, a rise in electrical transformation losses could cause energy consumption to increase more rapidly and could slow down or even reverse the decline in the ratio of energy consumption in industry to GDP.

There is no way to predict this effect quantitatively. If, however, the annual rate of increase in industrial energy consumption is arbitrarily increased to 4.0 percent (rather than the 3.06 percent called for in the four-year plan over the period 1980–85), and if real GDP is assumed to grow 7.0 percent a year from 1985 to 1990, the ratio of energy use to

27. Ibid.

Table 8-6. *Primary Energy Requirements in Taiwan, Central Case,*
1980 and 1990
Ten billions of kilocalories[a]

Sector	1980 (actual)	1990 (estimated)
Agriculture	947	1,557
Industry	16,548	23,401
Residential and commercial	5,044	9,587
Transportation	3,028	6,654
Nonenergy uses of fuels	3,040	5,704
Total	28,607	46,903

Sources: Transformation losses in making coal products and petroleum products have been distributed proportionately among all consuming sectors. The 1990 figures were derived by multiplying the estimated GDP in that year (2,022.6 billion 1976 New Taiwan dollars) by the sectoral ratios of energy use to GDP in table 8-5. Actual figures for 1980 are from Energy Committee, *Taiwan Energy Statistics, 1980*, p. 147. Numbers are rounded.
    a. Ten billion kilocalories approximately equal a thousand t.o.e.

GDP for industry in 1990 would be 115.7 kilocalories per 1976 New Taiwan dollar. This ratio was estimated earlier in table 8-5 to be 134.0 kilocalories per 1976 New Taiwan dollar in 1985. The annual rate of decline in the ratio from 1985 to 1990 would therefore be 2.9 percent, rather than the 4.0 percent rate of decline calculated for the period 1980–85 under assumptions of the four-year plan.

There are no compelling reasons to alter the annual rates of change in the ratios of energy use to GDP of the other energy-consuming sectors. Using those rates and the lower rate of decline for the industrial sector set forth above, overall energy intensity in 1990 can also be estimated (table 8-5). Substituting a slower rate of decline in the industry sector's ratio of energy use to GDP causes the overall energy intensity of Taiwan's economy to decline at an annual rate of 1.7 percent, compared with the 2.3 annual rate of decline from 1980 to 1985 that is implicit in the four-year plan.

The ratios of energy use to GDP for 1990 can be used to derive projections of total and sectoral energy consumption for that year (table 8-6). These projections will be taken as the central case definition of primary energy requirements in Taiwan for 1990. Actual energy consumption figures for 1980 are shown for comparison. The projections in Table 8-6 have total energy consumption increasing at an average annual rate of 5.07 percent from 1980 to 1990. GDP is assumed to grow at an average annual rate of 7.25 percent.[28] The implicit average GDP elasticity of demand is 0.699 during the 1980s.

This GDP elasticity can be used to obtain rough estimates of how much energy consumption would vary from the central case projections in table 8-6 if GDP grew more slowly or more rapidly than has been assumed. For illustrative purposes, case A assumes an average annual rate of GDP growth of 6.25 percent from 1980 to 1990; case B assumes an average annual growth rate of 8.25 percent from 1980 to 1990; energy consumption is in millions of t.o.e.:

	Case A	Central case	Case B
1980 (actual)	28.6	28.6	28.6
1990	43.8	46.9	50.1

Primary energy requirements in 1990 under case A would be 6.6 percent lower than those under the central case. Primary energy requirements under case B would be 6.8 percent higher than those under the central case in 1990.

Means of Meeting Taiwan's Future Energy Requirements

The four-year plan gives the following estimates (converted to thousands of t.o.e.) of how Taiwan's primary energy requirements in 1985 will be met: coal, 9,235; oil, 18,867; gas, 1,572; hydro, 1,281; nuclear, 5,752; for a total of 36,707.[29] These figures include energy used to generate electricity and to make coal products and petroleum products.

Table 8-7 presents an estimate of how Taiwan may meet its energy requirements in 1990. Because of the nature of the data, this estimate shows electricity separately. That is, the figures for coal and oil do not include the amounts of those fuels used to generate electricity. Those amounts will be estimated later in this section.

The derivation of the sectoral totals in table 8-7 was explained in the

28. GDP increased 5.5 percent in 1981. The four-year plan assumes an average annual rate of increase of 8.0 percent from 1981 to 1985. The average rate of increase for the period 1980–85 would therefore be 7.5 percent. An average annual increase in GDP of 7.0 percent from 1985 to 1990 was assumed in projecting energy consumption in 1990. The average rate of increase for the period 1980–90 would then be 7.25 percent.

29. CEPD, *Four-Year Plan*, table 8-1, p. 122. Estimated energy exports of 852 thousand t.o.e. have been subtracted from oil. Half of the estimated inventory increase of 979 thousand t.o.e. has been subtracted from coal; half has been subtracted from oil.

Table 8-7. *Estimated Sectoral Consumption of Coal, Oil, Gas,*
and Electricity in Taiwan, 1990
Ten billions of kilocalories[a]

Sector	Coal[b]	Oil[c]	Gas	Electricity[d]	Total[e]
Agriculture	*	1,311	0	246	1,557
Industry	5,961	5,962	986	10,492	23,401
Residential and commercial	403	2,761	489	5,934	9,587
Transportation	7	6,481	0	166	6,654
Nonenergy uses of fuels	0	4,923	781	0	5,704
Total	6,371	21,438	2,256	16,838	46,903

Source: See the text for derivation of sectoral totals. Numbers are rounded.
* Negligible.
  a. Ten billion kilocalories approximately equal a thousand t.o.e.
  b. Includes coal products.
  c. Includes petroleum products.
  d. Electricity and all other secondary fuels have been given the caloric values of the primary fuels required to make them (that is, transformation losses have not been subtracted).
  e. Sectoral totals for the agriculture, transportation, and nonenergy sectors have been distributed among the various fuels in the proportions that prevailed in 1979–80. The 1979–80 proportions were also used in the residential and commercial sector, but the share of electricity was increased by 5.5 percentage points (about 10 percent); the shares of other fuels were reduced proportionately. In the industrial sector the 1979–80 proportions were first applied to 22,412 ten billion kilocalories (what the sectoral total would be if industrial energy consumption increased at an annual rate of 3.0 percent from 1985 to 1990), with the following results (in ten billions of kilocalories): coal and oil, 11,923; gas, 986; and electricity, 9,503. The total for coal and oil was then divided about equally between those two fuels, and the difference between the estimated sectoral total of 23,401 and the hypothetical total of 22,412 was added to electricity. (See the text for justification of these changes.)

previous section of this chapter. The totals for the agriculture, transportation, and nonenergy sectors were distributed among coal, oil, gas, and electricity in the proportions that prevailed in 1979–80. The 1979–80 proportions were also used in the residential and commercial sector, but the share of electricity was increased by about 10 percent to allow for growth in the use of air conditioning. Nonenergy consumption was divided between oil and gas in the proportions projected for 1985 in the four-year plan.

In distributing the sectoral total for industry, two important trends were taken into account: the increased dependence on electricity accompanying the growth of technology-intensive industries, and the growth in the share of coal relative to oil in industrial energy consumption. The first of these trends was recognized by arbitrarily assuming that all of the projected acceleration in the annual rate of growth in industrial energy consumption (about 4.0 percent in 1985–90 compared with about 3.0 percent in 1980–85) would be in the form of electricity. The second trend was reflected by the equally arbitrary decision to give coal as large a share as oil in industrial energy consumption (not including coal and oil used to generate electricity).[30]

To estimate Taiwan's requirements for primary energy in 1990, the total for electricity in table 8-7 must be broken down into its components.[31] The first step is to separate electricity generated by hydropower and nuclear power from that made in conventional thermal plants. Projections for 1985 from the four-year plan are compared below with this study's estimates for 1990 (in thousands of t.o.e.):

	1985[32]	1990
Hydroelectric	1,281	1,473
Nuclear	5,752	5,752
Thermal	5,886	9,613
Total	12,919	16,838

Production of electricity by hydropower is estimated to increase 15 percent from 1985 to 1990 because the Taiwan Power Company plans to increase its hydroelectric capacity by that percentage over that period. Nuclear electricity output is assumed to be the same in 1990 as in 1985 because no new nuclear power plants will have been completed in that time.[33]

At the end of 1980, the Taiwan Power Company's thermal generating plants had a total capacity of 6,398 megawatts. One small 35-megawatt plant could burn only coal. Plants with a capacity of 945 megawatts could burn either coal or oil. The remaining plants, with a capacity of 5,418 megawatts, could burn only oil.[34] Approximately 85 percent of

---

30. From 1979 to 1980 the share of coal was 28.9 percent that of oil in the energy consumption of the industrial sector (plus the energy sector). See Energy Committee, *Taiwan Energy Statistics, 1980*, pp. 145, 147. In 1985 the consumption of coal is projected to be about 40 percent that of oil (excluding the use of these fuels to generate electricity). See CEPD, *Four-Year Plan*, p. 122. The share of coal relative to oil in the industrial sector would be even higher, since almost all coal is consumed in that sector, but substantial amounts of oil are used in other sectors.

31. In Taiwan's energy statistics, electricity is given the caloric value of the fuels that generate it, so that transformation losses need not be added. Electricity generated by hydropower and nuclear power is given the caloric value of equivalent amounts of electricity produced in thermal plants.

32. CEPD, *Four-Year Plan*, table 8-1, p. 122.

33. Taiwan Power Company, *Annual Report, 1981*, pp. 16–17.

34. Taiwan Power Company, "Business Opportunities with Taipower for American Businessmen," presentation to 1981 ROC-USA Trade and Investment Forum, Taipei, May 12–15, 1981, p. 2-3.

thermal generating capacity was therefore in oil-burning plants. Most of the remaining 15 percent was in plants that could burn either oil or coal.

By 1985 Taiwan Power will increase coal-burning generating capacity by 1,550 megawatts and oil-burning generating capacity by 799 megawatts. Three oil-burning plants with a combined capacity of 950 megawatts will be converted to burn coal. Taiwan Power will therefore have a total thermal generating capacity of 8,747 megawatts in 1985. Of this total, 5,307 megawatts must burn oil, and 3,440 megawatts will be in plants able to burn coal.[35] (It is not clear whether some of the new coal-burning plants could also use oil, nor is information available concerning the possible degradation of capacity in plants converted from oil to coal.)

Between 1985 and 1990 Taiwan Power plans to complete a 550-megawatt coal-burning plant and to convert a 500-megawatt oil-burning plant to coal. Total thermal generating capacity in 1990 will then be 9,247 megawatts, of which 4,490 megawatts (or 48.3 percent) will be in plants able to burn coal and 4,807 megawatts (or 51.7 percent) will still be in plants that have to burn oil.[36]

If the generation of electricity in coal-burning and oil-burning plants in 1990 is proportionate to capacity, the output of the former would have a caloric value of 4,643 thousand t.o.e., that of the latter 4,970 thousand t.o.e.[37] In 1980, however, Taiwan Power followed a policy of relying more on coal-burning plants than on oil-burning plants to hold down the consumption of expensive imported oil.[38] If this policy continues to be followed in 1990, coal-burning plants might account for 55 percent of thermal electricity output, oil-burning plants for 45 percent. The output of the coal-burning plants would then have a caloric value of 5,287 thousand t.o.e.

If these estimates of the use of coal and oil in generating electricity are combined with the estimates of the direct consumption of coal, oil, and gas in table 8-7, Taiwan's total requirements for hydrocarbons in 1990 (in thousands of t.o.e.) would be: coal, 11,658; oil, 25,764; gas,

35. See Taiwan Power Company, *Taipower and Its Development* (Taipei: Taipower, May 1981), p. 29; Taiwan Power Company, *Annual Report, 1981*, pp. 16–17; and *China Post*, April 7, 1981, p. 10.

36. Taiwan Power Company, *Annual Report, 1981*, pp. 16–17.

37. These are, of course, input values (that is, the caloric content of the fuels used to generate the electricity).

38. Figures provided by the Taiwan Power Company show that in 1980 coal-burning plants produced about 6.0 million kilowatt hours per megawatt of capacity, whereas oil-burning plants produced about 4.5 million kilowatt hours per megawatt.

2,256; for a total of 39,678. Domestic production could meet only a fraction of these requirements.

The four-year plan projects that domestic coal production will increase from 1,483 thousand t.o.e. in 1981 to 1,612 thousand t.o.e. in 1985, an annual rate of increase of 2.1 percent.[39] If coal production continues to increase at the same rate from 1985 to 1990—which is by no means assured—it would be 1,744 thousand t.o.e. in 1990. About 85 percent of estimated coal requirements in 1990, or 9,914 thousand t.o.e., would have to be imported.

Domestic oil production is very small and declining.[40] Barring new discoveries, oil import requirements would be roughly 25.6 million t.o.e. This estimate would have to be increased slightly if, as appears likely, domestic gas production in 1990 falls short of estimated requirements.

The four-year plan projects an annual rate of decline of 2.1 percent for natural gas production from 1981 to 1985 and estimates production in 1985 at 1,572 thousand t.o.e.[41] New discoveries of natural gas might be made, but if the rate of decline projected for the period of the four-year plan continues, production in 1990 would be 1,414 thousand t.o.e., or 63 percent of estimated total requirements. The shortfall of 842 thousand t.o.e., however, is far from being large enough to justify embarking on a program to import liquefied natural gas.[42] Rather, the answer would be to shift some gas consumers to oil. Requirements for imported oil would then be increased by about 800 thousand t.o.e., to a total of 26.4 million t.o.e.

Taiwan must import all of the uranium needed to operate its nuclear power plants. Consumption of U_3O_8 in 1980 (when nuclear generating capacity was 1,272 megawatts) was 262 short tons.[43] If consumption is proportionate to capacity, consumption in 1990 would be about 860 short

39. CEPD, *Four-Year Plan,* table 8-1, p. 122.

40. Ibid. The four-year plan estimates domestic oil production in 1985 at 153 thousand t.o.e.

41. Ibid.

42. Facilities to handle liquefied natural gas shipments should have a capacity of at least 2.5 million tons a year. (Conversation in Taipei, April 1982.) At 1,360 cubic meters per ton and 9,800 kilocalories per cubic meter, annual imports of liquefied natural gas would have to be at least 3.3 million t.o.e.

43. Taiwan Power Company, "Business Opportunities," p. 2-12. If the rule of thumb that every 1,000 megawatts (electric) of light water reactor capacity uses 142 metric tons of uranium is applied, requirements in 1990 would be 590 tons of uranium, or 767 short tons of U_3O_8.

tons (when nuclear generating capacity is projected to be 4,159 megawatts).

Estimates of Taiwan's energy import requirements in 1990 can be summarized as follows in original units (actual imports in 1980 are shown for comparison; coal and oil are in millions of metric tons; uranium is in millions of short tons of U_3O_8):[44]

	1980	1990
Coal[45]	4.6	14.6
Oil	20.8	26.4
Uranium	262.0	860.0

These projections of future import requirements, especially those for coal and oil, cannot be taken too literally. They rest on a series of assumptions that are unlikely to be precisely right. The cumulative effect of a number of small errors could be quite substantial. Nevertheless, the projections do indicate the likely direction of change. Coal imports will rise sharply during the 1980s, but increased use of coal and nuclear energy cannot keep oil import requirements from being higher in 1990 than they were in 1980.

Taiwan's coal and oil import requirements would be different from the projections if the rate of GDP growth is higher or lower than has been assumed in the central case. Coal and oil imports would not, however, necessarily be affected uniformly. Oil imports would probably be affected disproportionately by both faster and slower economic growth. Energy consumption in the transportation and nonenergy sectors, which are largely dependent on oil, is quite responsive to the rate of economic growth. In addition, in a period of slow growth Taiwan Power could be expected to reduce its reliance on oil-burning power plants. In a period of rapid growth all available generating capacity would be used more heavily, including that of oil-burning plants.

44. Energy Committee, *Taiwan Energy Statistics, 1980*, p. 92. Oil import figures in kiloliters have been converted to metric tons at the rate of 0.863 metric tons per kiloliter.

45. If the China Steel Corporation expands its facilities on schedule, and if it operates at capacity, requirements for imported coking coal would be 5.74 million metric tons in 1990. This assumes that 0.718 tons of coking coal are required to make a ton of steel (the average of 1979 and 1980 experience). See China Steel Corporation, *Annual Report, 1980* (Taipei: China Color Printing Co., 1980), pp. 2, 5. It further assumes that capacity will be 8.0 million tons in 1990 (see footnote 15, above). If coking coal is given a caloric value of 7,200 kilocalories per kilogram, coking coal imports would have a heat content of 4,136 thousand t.o.e. in 1990. Steam coal imports would then have a heat content of 5,778 thousand t.o.e. in 1990. Converting this figure at 6,500 kcal/kg, steam coal imports would be 8.89 million metric tons in 1990.

Taiwan now imports coal from the United States, Australia, South Africa, and Canada.[46] These countries will probably continue to be Taiwan's major sources of coal in the future. Importing coal from mainland China or from Siberia would be contrary to the anti-Communist policy of the government of the Republic of China.[47]

In 1980 Taiwan's imports of crude oil were distributed as follows (in percent): Kuwait, 49.2; Saudi Arabia, 34.1; Indonesia, 3.8; others, 13.0.[48] Prospects for Taiwan's diversifying these sources of oil imports are poor.

Taiwan imports all or most of its uranium from the United States.[49] The United States is also Taiwan's sole source of uranium enrichment services. The Republic of China has good relations with South Africa, and that country could in the future become a source of uranium and, possibly, of enrichment services.[50] Lack of diplomtic relations complicate, but do not necessarily preclude, Taiwan's obtaining uranium or enrichment services from other suppliers. A fifteen-year contract was recently signed with Rio Tinto Zinc Corporation of the United Kingdom to buy 4,400 tons of uranium, with delivery beginning in 1990.[51] If a contract to build a nuclear power plant on Taiwan were awarded to the French firm, Framatome, the deal might well include the supply of nuclear fuel by the French-controlled Eurodif.

Possible Developments after 1990

Taiwan's energy requirements may grow less rapidly in the 1990s than in the 1980s. As the economy matures, the rate of GDP growth may decline. Continued structural change in Taiwan's manufacturing industries may also bring a further decline in overall energy intensity. Neither of these developments, however, is certain. Well-chosen investments in technology-intensive industries with high-value-added products could

46. In 1980, 95.5 percent of Taiwan's coal imports were from the first three of these countries. See CEPD, *Four-Year Plan*, pp. 117–19.

47. It is possible, however, that Chinese Communist and Soviet authorities would see political, as well as economic, advantages in exporting coal to Taiwan.

48. CEPD, *Taiwan Statistical Data Book, 1981* (Taipei: CEPD, 1981), p. 216.

49. CEPD, *Four-Year Plan*, pp. 117–19.

50. Taiwan has reportedly bought small amounts of uranium from South Africa in the past.

51. *Asian Wall Street Journal Weekly*, December 20, 1982, p. 18.

produce a new upsurge in economic growth. Even though such industries have relatively low energy intensities, other changes could work to make overall energy intensity increase.

The spread of private automobiles and greater use of air conditioning could increase the ratios of energy use to GDP of the transportation sector and the residential and commercial sector. Increased reliance on electricity in factories, homes, and commercial facilities could produce large energy transformation losses.[52] The net effect of various possible developments on overall energy intensity cannot be foreseen. An end to the decline in energy intensity that has been projected for the 1980s is, however, a real possibility.

The displacement of oil by other sources of energy may continue in the 1990s, but perhaps at a slower rate. Several questions must be considered:

—How far will the nuclear power program be carried? The completion of nuclear plants 5 and 6 in 1984 and 1985 will temporarily check the rise in the use of hydrocarbons to generate electricity. When contracts will be let for plants 7 and 8 is uncertain.[53]

—Will the problem of matching refinery output with the demand for various petroleum products impose a constraint on substituting other fuels for the heavy fuel oil used to make electricity? Taiwan could find itself importing more crude oil to meet rising requirements for gasoline and kerosene and, as a result, producing more of the heavy refined products than can be used domestically. Whether this problem would be serious would depend on the export market for the surplus products.[54]

—Will liquefied natural gas be imported as an antipollution measure and as a means of diversifying sources of imported energy? If so, requirements for oil could be reduced, especially in the residential and commercial sector. Coal requirements could also be affected if liquefied natural gas took the place of electricity or if it were used to fuel electric power plants.

—Will relative costs continue to favor the substitution of other sources

52. To the extent that nuclear energy is used to produce more electricity, these transformation losses will be only hypothetical.

53. Because of reduced estimates of future demand for electricity, plans for these plants were indefinitely deferred in the summer of 1982. See *Asian Wall Street Journal Weekly*, July 19, 1982, p. 7.

54. The alternatives of importing gasoline and kerosene, rather than crude oil, would involve higher foreign exchange costs and the idling of some refining capacity.

of energy for oil? An affirmative answer is probably justified, although the price of coal has been rising.

—Finally, will political considerations set some limit to the decline in Taiwan's reliance on oil? Energy security for the Republic of China, which has formal ties with only a few countries, may require avoiding excessive dependence on any source of energy, including substitutes for oil. Moreover, good relations with Saudi Arabia have a value beyond the oil that it supplies but depend in part on the oil connection.

All of these uncertainties suggest that the continued rapid displacement of oil by other fuels is by no means inevitable. As in the case of energy intensity, no prediction can be made, but the long decline of dependence on oil could well slow down and might even come to an end.

CHAPTER NINE

Future Energy Developments in South Korea

THE WAYS in which South Korea obtains and uses energy are changing. The effort to substitute other sources of energy for oil should be increasingly successful, and continued economic modernization can be expected to alter the relative importance of various energy-consuming sectors.

Future Energy Requirements

Energy requirements depend on the size of the economy and its energy intensity. From 1973 to 1979 South Korea's gross national product (GNP) grew at an average annual rate of 9.7 percent.[1] A somewhat slower rate of growth is expected in the 1980s.[2] The energy intensity of the economy, measured by the ratio of primary energy supply to either gross domestic product (GDP) or GNP, fell fairly steadily during most of the 1970s (table 9-1). The ratio of energy supply to GNP, however, increased 5.3 percent in 1979 and 8.2 percent in 1980. These increases in

1. To make use of the GNP projections in Economic Planning Board, *Fifth Five-Year Economic-Social Development Plan, 1982–86* (Seoul: EPB, 1981; in Korean), GNP rather than GDP will be used in this chapter. GNP differs from GDP by the amount of net factor income paid or earned abroad.

2. Real GNP fell 6.2 percent in 1980 but increased 7.1 percent in 1981 and 5.4 percent in 1982. The Korea Development Institute estimates 8.8 percent growth in 1983 and forecasts 8.1 percent growth in 1984. See *Korea Herald*, April 6, 1983, p. 1, and October 6, 1983, p. 7.

Table 9-1. *Overall Energy Intensity of the Economy of South Korea, 1973–80*

Year	Gross national product (billions of 1980 won)	Primary energy supply (ten billions of kilocalories)[a]	Energy intensity (kilocalories per 1980 won)
1973	20,984	25,282	12.05
1974	22,664	25,501	11.25
1975	24,280	27,067	11.15
1976	27,956	29,796	10.66
1977	30,824	33,067	10.73
1978	34,407	36,136	10.50
1979	36,593	40,475	11.06
1980	34,322	41,078	11.97

Sources: GNP: 1973–79 (in 1975 won), Economic Planning Board, *Korea Statistical Yearbook, 1981* (Seoul: EPB, December 1981), pp. 464–65; 1980 (in 1980 won), Bank of Korea, *Monthly Economic Statistics*, vol. 35 (December 1981), p. 132. All GNP figures in this table are in 1980 won. Primary energy supply: 1973–79, Korea Energy Research Institute, *Basic Data for Energy Policy Study* (Seoul: KERI, June 1980; in Korean), table I-1; 1980, Ministry of Energy and Resources, *Five-Year Energy Plan, 1982–86*, (Seoul: The Ministry, December 1981; in Korean), appendix, p. 4. Sources assumed 34.4 percent generating efficiency for hydroelectric and nuclear power; this table assumes 35.1 percent. Nonenergy consumption of petroleum products was subtracted from source's 1980 total for primary energy to make it comparable with 1973–79 figures. Numbers are rounded.
a. Ten billion kilocalories approximately equal a thousand metric tons of oil equivalent (t.o.e.).

energy intensity could be the beginning of a new trend, but it is more likely that they were the transitory effect of the 1979–80 oil crisis, aggravated by the political instability following the assassination of President Park Chung Hee in October 1979.

The increase in overall energy intensity in 1979 and 1980 resulted from the following differences, based on table 9-1, in annual rates of change in GNP and primary energy supply:

	GNP	Primary energy supply
1978	11.6	9.3
1979	6.4	12.0
1980	−6.2	1.5

In 1979 the increase in energy intensity was caused by opposite movements in the rates of change in GNP and in primary energy supply. In 1980 it was caused by the difference in the rates of decline of these two variables.

In both 1979 and 1980 the industrial sector was the most volatile component of total energy use (which equals primary energy supply). In 1979 industrial energy consumption rose 26.3 percent, and in 1980 it fell 11.1 percent. The sharp rise in industrial energy consumption in 1979, at a time when economic growth was slowing down, may have been caused by the completion of energy-intensive production facilities. The drop in industrial energy consumption in 1980 was the result of the depressed state of the economy in that year.

The Five-Year Plan

The five-year plan for the period 1982–86[3] is comparable in scope and quality to the four-year plan for Taiwan that was used in chapter 8. The five-year plan's original estimates of energy use in 1986 will be adopted in this chapter as the central case projections for that year. Subsequent modifications of those estimates, as reported in the press, will be cited when appropriate.[4]

The five-year plan assumes that real GNP will increase at an average annual rate of 7.6 percent from 1981 to 1986. The assumption made about real energy prices is not clear. A summary draft of the plan includes the statement that "the oil price is expected to show a rising trend."[5] This draft also states that during the plan period the GNP deflator will rise at an annual rate of 10.8 percent.[6] Quite substantial increases in nominal oil prices would therefore have to occur to cause an increase in real energy prices.

The five-year plan projects a 7.0 percent annual rate of growth for total primary energy requirements (or total energy use) from 1981 to 1986. This projection is substantially below the 10 percent annual rate of growth in energy consumption estimated in 1978 by the Korea Development Institute for the period 1980–91. The institute, however, assumed a 10 percent annual increase in GDP.[7]

3. EPB, *Fifth Five-Year Economic-Social Development Plan*. This overall plan is supported by a detailed energy plan, Ministry of Energy and Resources, *Five-Year Energy Plan, 1982–86* (Seoul: The Ministry, December 1981; in Korean). The overall plan will be cited hereafter as the *Five-Year Plan*. The supporting energy plan will be cited hereafter as the *Five-Year Energy Plan*.

4. Only the highlights of the modified estimates have been reported. See *Korea Herald*, August 19, 1982, p. 6, and September 25, 1982, p. 6.

5. EPB, *A Summary Draft of the Fifth Five-Year Economic and Social Development Plan, 1982–86* (Seoul: EPB, August 1981; in English), p. 91. This document will be cited hereafter as *Summary Draft of Five-Year Plan*.

6. Ibid., p. 24.

7. For an analysis of the differences between the five-year plan's projection and several other projections, including that of the Korea Development Institute, see Joy Dunkerley and others, "Future Energy Consumption in India, Brazil, Republic of Korea, and Mexico," RFF Discussion Paper (Washington, D.C.: Resources for the Future, March 1982), pp. 60–78.

Table 9-2. *Primary Energy Requirements in South Korea under the Five-Year Plan, 1981–86*
Ten billions of kilocalories[a]

| | 1981 | | 1986 | | Annual rate of change |
Sector	Quantity	Percent	Quantity	Percent	(percent)
Industrial[b]	14,225	29.8	16,909	25.2	3.5
Residential and commercial	14,577	30.5	17,541	26.2	3.8
Transportation	5,970	12.5	11,021	16.4	13.0
Public and other	2,136	4.5	2,893	4.3	6.3
Nonenergy uses of fuels[b]	4,329	9.1	6,585	9.8	8.8
Electricity generation losses[c]	6,525	13.7	12,120	18.1	13.2
Total	47,762	100.0	67,069	100.0	7.0

Source: Ministry of Energy and Resources, *Five-Year Energy Plan, 1982–86* (Seoul: The Ministry, December 1981; in Korean), appendix, pp. 3, 4, 6, 8, 13, 16 (hereafter *Five-Year Energy Plan*). Figures for various fuels in original units were converted to t.o.e. or kilocalories as follows. Petroleum products: 1981, 0.148 t.o.e. per barrel; 1986, 0.146 t.o.e. per barrel. Liquefied petroleum gas: 1.187 t.o.e. per ton. Liquefied natural gas, 1.30 t.o.e. per ton. Numbers are rounded.
a. Ten billion kilocalories approximately equal a thousand t.o.e.
b. Nonenergy use of coal products was assumed to be of metallurgical coke and was shifted from the nonenergy sector to the industrial sector.
c. Electricity generation losses are the difference between total inputs to electricity generation and consumption. Total inputs to electricity generation are the difference between primary energy requirements and energy consumption in the consuming sectors. Hydroelectric and nuclear inputs are equivalent to those of thermal generating plants operating at 34.4 percent efficiency.

Table 9-2 gives sectoral energy requirements in 1981 and a projection of sectoral requirements in 1986. Both were derived from the five-year energy plan. Sectoral requirements include transformation losses incurred in making coal products and petroleum products. Transformation losses in generating electricity are shown separately.

Total primary energy requirements in the five-year energy plan (and in table 9-2) are not fully comparable with total requirements in the energy balance tables for Korea in chapter 3. In the past, Korean energy statistics have not included nonenergy uses of fuels (unless coking coal is so regarded). The five-year energy plan, however, explicitly includes nonenergy uses of petroleum products, coal products, and gas (presumably liquefied petroleum gas).[8] In table 9-2 nonenergy use of coal products (presumably of metallurgical coke) has been shifted to the industrial sector.

Energy consumption in the industrial sector and in the residential and commercial sector is projected to increase at annual rates of only 3.5

8. In the past, Korean energy statistics did not show gas separately. Liquefied petroleum gas was probably included in petroleum products, and imports of liquefied natural gas are not scheduled to begin until the mid-1980s.

percent and 3.8 percent, respectively. The amounts of energy consumed in the transportation sector and lost in generating electricity are projected to rise at annual rates of 13.0 percent and 13.2 percent, respectively. Energy use in the other two sectors, nonenergy uses and public and other, is projected to increase at annual rates closer to the 7.0 percent projected for total energy use. The wide differences in sectoral growth rates would cause marked declines in the shares of the industrial sector and the residential and commercial sector in total energy use. The shares of transportation and electricity generation losses would increase correspondingly.

The changes in the structure of energy demand projected in the five-year plan and set forth in table 9-2 are probably attributable in part to structural changes in the Korean economy. This is difficult to demonstrate, however, because—with one exception—energy-consuming sectors and the categories into which GNP is divided do not match. The exception is the industrial sector in energy use and the manufacturing sector in the breakdown of the industrial origins of GNP. The five-year plan has manufacturing increasing from 29 percent of GNP in 1981 to 34 percent in 1986.[9] This contrasts with the plan's projection of a decrease in the industrial sector's share of energy use—from 29.8 percent in 1981 to 25.2 percent in 1986.

The plan clearly anticipates a significant decrease in the energy intensity of Korean manufacturing industries. This decrease is to be brought about by discouraging the construction or expansion of "high-energy-consuming factories," encouraging the replacement of old and relatively inefficient facilities, enforcing energy standards, and improving the technology for energy management.[10] The allocation of fixed investment funds within manufacturing industry also should help to reduce energy intensity. The composition of fixed investment in manufacturing called for in the five-year plan is compared below with the actual percentage composition in the period 1972–79:[11]

	1972–79	1982–86
Light industry	37.3	32.9
Chemical industry	19.3	20.9
Metal industry	19.2	9.0
Machinery industry	24.2	37.2

9. *Five-Year Plan*, pp. 138–39.
10. Ibid., pp. 67–74.
11. *Summary Draft of Five-Year Plan*, p. 38.

The principal changes projected in the allocation of fixed investment are the large decrease in the share of the metal industry, which is relatively energy intensive, and the large increase in the share of the machinery industry, which is much less energy intensive. The machinery industry includes electronics and transportation equipment. The share of the chemical industry would have been lower in 1982–86 but for the need to invest in desulfurizing facilities.

The industrial development strategy of the Korean government during the period of the fifth five-year plan "will be based more on market mechanisms and less on government intervention." The role of the government in promoting so-called strategic industries will be reduced; "investment decisions will be left more to the private sector while the government will seek only to provide basic guidelines."[12] Competition, including foreign competition, will be encouraged.

As noted above, the five-year plan assumed that GNP will grow at an annual average rate of 7.6 percent from 1982 to 1986. GNP in 1986 is projected to be 53,676.9 billion 1980 won.[13] Actual GNP in 1981 was 36,758 billion 1980 won, 1.2 percent below the 37,216 billion 1980 won estimated for that year in the plan.[14] To reach the GNP projected for 1986, the annual rate of growth from 1981 to 1986 would have to be 7.9 percent. Whether this rate of growth can be achieved depends to a large degree on the extent to which Korea's export markets recover from the recession of the early 1980s.

By the five-year plan's estimates of GNP and total energy use in 1981 and 1986, the overall energy intensity of the Korean economy would be 12.83 kilocalories per won in 1981 and 12.49 kilocalories per won in 1986.[15] The projected decline in intensity from 1981 to 1986 is quite modest—only 2.65 percent, or an annual rate of decline of about 0.5 percent.

This small decline in overall energy intensity is the net result of the quite different changes in sectoral ratios of energy use to GNP (table 9-3). The figures in table 9-3 indicate that the five-year plan projects no

12. Ibid., especially pp. 75–77.
13. *Five-Year Plan*, pp. 138–39.
14. GNP in 1980 at current market prices was 34,321.6 billion won. See Bank of Korea, *Monthly Economic Statistics*, vol. 35 (December 1981), p. 132. GNP reportedly grew 7.1 percent in 1981. See *Korea Herald*, May 14, 1982, p. 1.
15. These energy intensities are not comparable with those in table 9-1 because they take account of nonenergy uses of petroleum products that were not recorded in past Korean energy statistics.

Table 9-3. *Sectoral Ratios of Energy Use to GNP in South Korea,*
Estimates for 1981 and 1986 and Illustrative Projections for 1991
Kilocalories per 1980 won

Sector	Energy use/GNP			Annual rate of change (percent)	
	1981	*1986*	*1991*	*1981–86*	*1986–91*
Industrial	3.82	3.15	2.57	−3.8	−4.0
Residential and commercial	3.92	3.27	2.96	−3.6	−2.0
Transportation	1.60	2.05	2.49	5.1	4.0
Public and other	0.57	0.54	0.51	−1.1	−1.1
Nonenergy uses of fuels	1.16	1.23	1.26	1.2	0.5
Electricity generation losses	1.75	2.26	2.62	5.2	3.0
Total	12.83	12.49	12.41	−0.5	−0.1

Sources: Ratios of energy use to GNP in 1981 and 1986 were derived from the *Five-Year Energy Plan*. Ratios for 1991 are illustrative and are based on judgments about possible sectoral developments after 1986 (see the text). Numbers are rounded.

general decline in ratios of energy use to GNP. Instead, about 73 percent of the substantial decreases in the ratios for the industrial and residential and commercial sectors is counterbalanced by increases in those for the transportation sector and for electricity generation losses.

Energy Requirements after 1986

The rates of growth of both the Korean economy and its energy intensity may differ in the period of the sixth five-year plan (1987–91) from the rates projected in the current plan (which ends in 1986). A somewhat slower rate of growth of GNP—say, 7.0 percent annually— appears to be a reasonable assumption.[16] Energy intensity will be influenced both by general factors and by factors affecting particular energy-consuming sectors.

The effect of general factors on energy intensity is not clear. As pointed out, the assumptions of the current five-year plan about real energy prices are uncertain. The consequences of assuming constant real energy prices after 1986, as is done in this study's central case projections, therefore cannot be estimated. Thermal efficiency will probably increase somewhat during the period of the current plan as a result of a shift to more efficient fuels. (See table 9-6.) The continued

16. In a table of major indicators for the period 1980–91, the English summary of the fifth five-year plan shows a 7.0 percent annual growth of GNP for the period 1987– 91. See *Summary of Five-Year Plan*, p. 151.

electrification of the Korean economy may bring further gains in thermal efficiency after 1986, but the rate of improvement could slow down as the gains from phasing out noncommercial fuels are exhausted. Net imports of embodied energy, which reduce overall energy intensity, will decrease if the reduction in the trade deficit called for in the current five-year plan is achieved.[17] Whether the trade deficit will continue to shrink after 1986—and, if so, at what rate—can only be a matter for speculation.

Somewhat more useful results can be obtained by asking whether the rates of change in sectoral ratios of energy use to GNP implicitly projected in the five-year plan (and given in table 9-3) are likely to increase or decrease after 1986.

The continued decline in the ratio of energy use to GNP in the industrial sector depends on whether improvements in energy efficiency keep ahead of the increase in the importance of industry in the economy. At some point the economic importance of the industrial sector should slow down, and some of the measures adopted to improve energy efficiency during the current five-year plan (replacing relatively inefficient facilities, improving energy management technology, and favoring less energy-intensive activities in the allocation of fixed investment funds) should have continuing, and possibly increasing, effects after 1986. The ratio of energy use to GNP in the industrial sector could therefore continue to fall at the same rate, or possibly a bit faster.

Barring significant changes in the relationship between GNP and disposable income—and there are no reasons to anticipate any—changes in the ratio of energy use to GNP of the residential and commercial sector will depend on trends in energy efficiency. A large part of the gains in thermal efficiency to be realized by phasing out noncommercial fuels (which are used almost exclusively in this sector) will have been achieved by 1986. In addition, increasing family incomes will stimulate the acquisition of energy-using home appliances. The rate of decline in this sector's ratio of energy use to GNP may therefore slow down considerably after 1986.

The projected rise in the ratio of energy use to GNP of the transpor-

---

17. Ibid., p. 23. The plan projects a decrease in the trade deficit (in current U.S. dollars) from $4.2 billion in 1981 to $2.5 billion in 1986. Because fuel imports (oil, coal, and gas) are projected to increase over the period of the plan, all of the decrease in the trade deficit would be reflected in net nonfuel imports. Changes in the relative energy intensities of imports and exports would not be able to overcome the effect of the decrease in the trade deficit on net imports of embodied energy.

tation sector from 1981 to 1986 is probably attributable to a shift from rail to less efficient highway transportation. Efforts begun during the current five-year plan to promote the use of railways and coastal shipping could yield increasing results after 1986 and could reduce the rate of increase in the transportation sector's ratio of energy use to GNP.

Generating efficiency in Korean electric power plants is already high and is not likely to improve significantly. Losses in generating electricity will therefore vary directly with electricity consumption. The Korea Electric Power Corporation expects electricity consumption to grow at an annual rate of 11.1 percent from 1987 to 1991.[18] (The five-year plan projects an annual rate of increase of 10.7 percent from 1981 to 1986.) The corporation assumes, however, that GNP will grow 8.0 percent a year from 1987 to 1991. If the 7.0 percent GNP growth rate assumed in the present study were to be used, the corporation's projection would have to be scaled down.[19] The rate of increase in the ratio of electricity generation losses to GNP would therefore decline.

Competition from foreign producers with access to cheaper feedstocks could check the growth of the Korean petrochemical industry and could slow down, or even stop, the increase in the ratio of energy use to GNP of the nonenergy sector. Too little is known about the small public and other sector to make judgments about the future rate of change in its ratio of energy use to GNP, so the rate of change in this ratio will be assumed to remain about the same after 1986 as it is projected to be in the period 1981–86.

The above judgments about rates of change in sectoral ratios of energy use to GNP cannot easily be expressed quantitatively. Table 9-3, however, presents an illustrative case consistent with these judgments and derives sectoral ratios for 1991, the last year of the sixth five-year economic planning period.

The case for 1991 cannot be taken too literally. Its main utility is to show that under plausible assumptions the moderate decline in the energy intensity of the Korean economy projected by the current five-year plan could slow down and virtually cease in the next planning

18. Korea Electric Power Corporation, *Electric Power in Korea, 1982* (Seoul: KEPCO, 1982), p. 14.

19. The implicit GNP elasticity of demand for electricity in the estimates of the five-year plan for the period 1981–86 is about 1.4. If GNP grows 8.0 percent a year, electricity requirements would therefore increase 11.2 percent annually. If the GNP growth rate is 7.0 percent, the electricity growth rate would be 9.8 percent.

Table 9-4. *Sectoral Requirements of Primary Energy
in South Korea, Estimates for 1981 and 1986
and Illustrative Projections for 1991*
Ten trillions of kilocalories[b]

Sector	1981	1986	1991
Industrial	14.2	16.9	19.3
Residential and commercial	14.6	17.5	22.3
Transportation	6.0	11.0	18.7
Public and other	2.1	2.9	3.8
Nonenergy uses of fuels	4.3	6.6	9.5
Electricity generation losses	6.5	12.1	19.7
Total	47.8	67.1	93.3

Sources: Tables 9-2 and 9-3. Numbers are rounded.
a. Ten trillion kilocalories approximately equal a million t.o.e.

period. In the illustrative case, declines in the ratios of energy use to GNP of the industrial sector and the residential and commercial sector are almost exactly balanced by increases in the ratios of the transportation sector and the electricity generation sector.

If GNP grows 7.0 percent annually from 1986 to 1991,[20] the ratios of energy use to GNP for 1991 presented in table 9-3 would imply the requirements for primary energy shown in table 9-4. Estimates for 1981 and 1986 from table 9-2 are shown for comparison. Under this illustrative projection, total requirements for primary energy would increase at an annual rate of 6.8 percent from 1986 to 1991. As previously noted, under the five-year plan primary energy requirements are projected to increase 7.0 percent annually from 1981 to 1986.

Alternatives to the Central Case

The implicit GNP elasticity of demand for energy in the central case is 0.92 for the period 1981–86 and 0.97 for the period 1986–91. With these elasticities, rough estimates can be made of energy requirements under assumptions about the rate of GNP growth that are higher or lower than those of the central case. Case A assumes that the annual rate of increase of GNP will be 6.6 percent from 1981 to 1986 and 6.0 percent from 1986 to 1991. Case B assumes annual rates of increase of GNP of 8.6 percent

20. The five-year plan estimates that GNP will be 53,677 billion 1980 won in 1986. If GNP grows at an annual rate of 7.0 percent, it would be 75,285 billion 1980 won in 1991.

from 1981 to 1986 and of 8.0 percent from 1986 to 1991. (These alternative assumptions are one percentage point below and one percentage point above those of the central case.)

Primary energy requirements under cases A and B and the central case would be (in millions of metric tons of oil equivalent, t.o.e.):

	Case A	*Central case*	*Case B*
1981	47.8	47.8	47.8
1986	64.2	67.1	69.9
1991	87.5	93.3	101.6

Under case A, primary energy requirements would be 4.3 percent lower than those under the central case in 1986 and 6.2 percent lower in 1991. Under case B, requirements would be 4.2 percent higher than those under the central case in 1986 and 8.9 percent higher in 1991.

The modified energy estimates announced by the Ministry of Energy and Resources in September 1982 project a 6.1 percent annual increase in total energy use during the period of the five-year plan. Primary energy requirements in 1986 would then be 64.3 million t.o.e. The similarity of this estimate to case A above is a coincidence. The modified official estimates use a higher annual economic growth rate (7.2 percent) than in case A but a lower income elasticity (0.84).

Means of Meeting South Korea's Future Energy Requirements

Table 9-5 presents estimates of sectoral energy consumption in 1981 and 1986 by fuel. These estimates were derived from the five-year plan and the supporting five-year energy plan.[21] In table 9-6 these estimates have been converted to percentages in order to bring out more clearly the structural changes that the plan projects for the period 1981–86.

Oil will continue to be the only fuel used in the nonenergy sector,[22] and it will continue to enjoy a near monopoly in the transportation sector. The already high share of oil in the public and other sector will increase at the expense of coal. The most interesting changes in the

21. These estimates are of final energy consumption plus the transformation losses incurred in making petroleum products and coal products.

22. The five-year plan regards coking coal as a nonenergy material. To facilitate comparison with other countries, coking coal in this study has been included in the energy consumption of the industrial sector.

Table 9-5. *Estimated Sectoral Consumption of Oil, Coal, Gas, Electricity, and Noncommercial Fuels in South Korea, 1981 and 1986*
Ten billions of kilocalories[a]

Sector	1981	1986	Annual rate of change (percent)
Industrial			
Oil[b]	6,818	8,027	3.32
Coal[c]	5,255	5,342	0.33
Electricity	2,152	3,540	10.47
Total	14,225	16,909	3.52
Residential and commercial			
Oil[b]	2,407	3,943	10.37
Coal[c]	9,074	10,818	3.58
Gas[d]	0	152	n.a.
Noncommercial[e]	2,442	1,501	−9.27
Electricity	654	1,127	11.50
Total	14,577	17,541	3.77
Transportation			
Oil[b]	5,931	10,935	13.02
Electricity	39	86	17.14
Total	5,970	11,021	13.04
Public and other			
Oil[b]	1,825	2,550	6.92
Coal[c]	96	11	−35.16
Electricity	215	332	9.08
Total	2,136	2,893	6.26
Nonenergy uses of fuels			
Oil[b]	4,329	6,585	8.75
Total sectoral consumption			
Oil[b]	21,310	32,040	8.50
Coal[c]	14,425	16,171	2.31
Gas[d]	0	152	n.a.
Noncommercial[e]	2,442	1,501	−9.27
Electricity	3,060	5,085	10.69
Total	41,237	54,949	5.91

Source: See table 9-2. Numbers are rounded.
n.a. Not available.
a. Ten billion kilocalories approximately equal a thousand t.o.e.
b. Oil includes petroleum products and transformation losses in making them (including liquefied petroleum gas).
c. Coal includes coal products and transformation losses in making them.
d. Gas is liquefied natural gas.
e. Noncommercial fuels are wood, charcoal, and crop residues.

Table 9-6. *Estimated Structure of Sectoral Energy Consumption in South Korea, 1981 and 1986*
Percent

Fuel	Industrial		Residential and commercial		Transpor- tation		Public and other		Nonenergy uses of fuels		Total	
	1981	1986	1981	1986	1981	1986	1981	1986	1981	1986	1981	1986
Oil	47.9	47.5	16.5	22.5	99.3	99.2	85.4	88.1	100.0	100.0	51.8	58.3
Coal	36.9	31.6	62.2	61.7	0.0	0.0	4.5	0.4	0.0	0.0	35.0	29.4
Gas	0.0	0.0	0.0	0.9	0.0	0.0	0.0	0.0	0.0	0.0	0.0	0.3
Noncommercial	0.0	0.0	16.8	8.5	0.0	0.0	0.0	0.0	0.0	0.0	5.9	2.7
Electricity	15.1	20.9	4.5	6.4	0.6	0.8	10.1	11.5	0.0	0.0	7.4	9.2
Total	100.0	100.0	100.0	100.0	100.0	100.0	100.0	100.0	100.0	100.0	100.0	100.0

Source: Table 9-5. Numbers are rounded.

structure of energy supply are, however, projected to occur in the industrial sector and in the residential and commercial sector.

The share of electricity in industrial energy consumption is expected to increase from 15.1 percent in 1981 to 20.9 percent in 1986. This increase in electricity use will principally displace coal as an industrial fuel; oil will be only marginally affected.

In the residential and commercial sector, the major change projected in the five-year plan is a sharp drop in the share of noncommercial fuels, from 16.8 percent in 1981 to 8.5 percent in 1986. Most of this decrease is to be made up by a rise in the share of oil, but the share of electricity is also expected to increase. In addition, a small share of total sectoral consumption is planned to go to imported liquefied natural gas.[23] The share of coal is projected to decline only slightly.

The combined effect of projected sectoral changes will be to increase the share of oil by 6.5 percentage points and to decrease the share of coal by 5.6 percentage points. The share of noncommercial fuels will fall by 3.0 percentage points, and the share of electricity will rise by 1.8 percentage points. Liquefied natural gas, scheduled to be imported beginning in 1986, is expected to provide 0.3 percent of total sectoral energy consumption in 1986.

This picture changes appreciably if the energy lost in generating electricity is taken into account (table 9-7). These estimates have been expressed as percentages in table 9-8.

23. After prolonged negotiations between the Korean and Indonesian governments, the import program for liquefied natural gas is now scheduled to begin in 1986. See *Korea Herald,* May 12, 1983, p. 6.

Table 9-7. *Estimated Requirements of Various Forms of Primary Energy in South Korea, 1981 and 1986*
Ten billions of kilocalories[a]

	1981			1986		
Fuel	Electricity generation	Other	Total	Electricity generation	Other	Total
Oil	7,383	21,310	28,693	3,219	32,041	35,260
Coal	869	14,425	15,294	4,947	16,170	21,117
Gas	0	0	0	1,752	152	1,904
Noncommercial	0	2,442	2,442	0	1,501	1,501
Hydroelectric	559	0	559	726	0	726
Nuclear	774	0	774	6,561	0	6,561
Total	9,585	38,177	47,762	17,205	49,864	67,069

Sources: Tables 9-2 and 9-5. Numbers are rounded.
a. Ten billion kilocalories approximately equal a thousand t.o.e.
b. Hydroelectric and nuclear inputs were calculated on an assumption of 34.4 percent generating efficiency (as in the *Five-Year Energy Plan*).

Table 9-8. *Estimated Structure of Primary Energy Supply in South Korea, 1981 and 1986*
Percent

	1981			1986		
Fuel	Electricity generation	Other	Total	Electricity generation	Other	Total
Oil	77.0	55.8	60.1	18.7	64.3	52.6
Coal	9.1	37.8	32.0	28.8	32.4	31.5
Gas	0.0	0.0	0.0	10.2	0.3	2.8
Noncommercial	0.0	6.4	5.1	0.0	3.0	2.2
Hydroelectric	5.8	0.0	1.2	4.2	0.0	1.1
Nuclear	8.1	0.0	1.6	38.1	0.0	9.8
Total	100.0	100.0	100.0	100.0	100.0	100.0

Source: Table 9-7. Numbers are rounded.

Table 9-8 shows that the changes in the structure of primary energy supply during the period 1981–86 will be caused principally by dramatic shifts in the ways in which electricity is generated. Impressive increases in the shares of nuclear power and coal in the energy used to make electricity make possible a sharp drop in reliance on oil. All of the projected success in reducing dependence on oil from 60.1 percent of total primary energy in 1981 to 52.6 percent in 1986 can be credited to

these changes in the generation of electricity. Without these changes—
as tables 9-6, 9-7, and 9-8 show—dependence on oil would actually
increase.

The modified energy estimates for 1986 announced in September 1982
reduce the share of oil in total energy use to 46.2 percent and defer the
availability of liquefied natural gas to the end of 1986. The share of
nuclear power is raised slightly to over 10 percent. The shares of
noncommercial fuels and hydropower are presumably unchanged. To
make up for the reduction in the share of oil and deferred imports of
liquefied natural gas, the share of coal would have to increase to about
40 percent. How this could be done is not explained in published reports.

Table 9-9 presents rough estimates of the sectoral consumption of
various fuels in 1991. Estimates for 1986, based on the current five-year
plan, are also shown for comparison. The sectoral totals for 1991 are
from the illustrative projection developed earlier (table 9-4). The Korea
Electric Power Corporation has published estimates of the sectoral use
of electricity and liquefied natural gas in 1991.[24] With minor modifications, these estimates have been used in table 9-9.

In table 9-9 total electricity consumption (after transmission and
distribution losses) was estimated to be 8.4 million t.o.e.[25] The Korea
Electric Power Corporation estimated that in 1991 67.5 percent of
electricity sales will be to industrial consumers and 32.5 percent to
residential and commercial consumers. Nothing was said about sales in
the transportation sector or in the public and other sector, but it has been
assumed in table 9-9 that the former are included in the industrial sector
and the latter in the residential and commercial sector. It has further
been assumed that the share of electricity in the energy consumption of
both the transportation sector and the public and other sector will be
only slightly greater in 1991 than the share projected in the five-year plan
for 1986. This assumption would have to be revised for the transportation
sector if significant railway mileage were electrified before 1991.

24. Korea Electric Power Corporation, *Electric Power*, pp. 14, 26. The Korean
government has placed the corporation in charge of the Liquefied Natural Gas Intro-
duction Project.

25. This estimate uses the ratio of electricity generation losses to GNP from table
9-3. It is assumed that GNP in 1991 will be 75,285 billion 1980 won (based on a 7.0
percent annual rate of growth from 1986 to 1991) and that electricity transformation
losses (including transmission and distribution losses) will be 70 percent of the total
energy used in making electricity (as they are projected to be in 1986 in the current
five-year plan).

Table 9-9. *Estimated Sectoral Consumption of Various Fuels
in South Korea, 1986 and 1991*
Ten trillions of kilocalories[a]

Sector	1986	1991	Annual rate of change (percent)
Industrial			
Oil	8.0	8.4	1.0
Coal	5.3	5.4	0.4
Electricity	3.5	5.5	9.5
Total	16.9	19.3	2.7
Residential and commercial			
Oil	3.9	7.5	14.0
Coal	10.8	10.8	0.0
Gas	0.2	0.8	32.0
Noncommercial	1.5	0.9	−9.7
Electricity	1.1	2.3	15.9
Total	17.5	22.3	5.0
Transportation			
Oil	10.9	18.5	11.2
Electricity	0.1	0.2	14.9
Total	11.0	18.7	11.2
Public and other			
Oil	2.6	3.4	5.5
Coal	*	0.0	0.0
Electricity	0.3	0.4	5.9
Total	2.9	3.8	5.6
Nonenergy uses of fuels			
Oil	6.6	9.5	7.6
Total sectoral consumption			
Oil	32.0	47.3	8.1
Coal	16.1	16.2	0.1
Gas	0.2	0.8	32.0
Noncommercial	1.5	0.9	−9.7
Electricity	5.1	8.4	10.5
Total	54.9	73.6	6.0

Sources: 1986 figures are from table 9-5. See the text for explanation of 1991 figures. Numbers are rounded.
* Negligible.
a. Ten trillion kilocalories approximately equal a million t.o.e.

The corporation also estimated that 79.7 percent of the 3.0 million tons of liquefied natural gas to be imported in 1991 will be used to generate electricity, and that the remaining 20.3 percent will be distributed as town gas to residential and commercial consumers. At 1.3 t.o.e. per metric ton, the 0.61 million tons of liquefied natural gas going to the residential and commercial sector would have an energy content of about 0.8 million t.o.e.

The nonenergy sector can be assumed to rely entirely on oil, and, except for small amounts of electricity, oil will probably continue to be the source of energy in both the transportation sector and the public and other sector. Oil can be treated as the residual fuel in the other two sectors, but estimating the future consumption of coal and noncommercial fuels in those sectors is difficult.

Despite the government's policy of encouraging the substitution of coal for oil wherever possible, the current five-year plan has the use of oil in industry expanding about ten times as rapidly as the use of coal. Table 9-9 assumes that in the period 1986–91 the industrial use of coal will continue to increase by only about 0.3 percent annually.[26] The use of oil, the residual fuel, would then increase by about 1.0 percent a year, compared with 3.3 percent a year from 1981 to 1986. The slowdown in the use of oil is explained by the reduced rate of growth of sectoral energy consumption and by the continued rapid electrification of Korean industry.

With respect to the residential and commercial sector, the rapid phasing-out of noncommercial fuels can be expected to continue. The growth in the use of coal briquets for space heating may well cease as consumers turn to fuels more suited to use in multistory apartment houses and office buildings. Under government plans imported liquefied natural gas will eventually take over a large part of space heating, but this process will still be in its early stages in the period 1986–91. For a time oil, and to a lesser extent electricity, must satisfy the increasing energy needs of the residential and commercial sector.

The projections of sectoral energy consumption in 1991 that are presented in table 9-9 rest on a number of arbitrary assumptions. These projections are in no sense forecasts. They do suggest, however, that

---

26. A higher rate of growth would be justified if it appeared that the second integrated steel mill will actually be completed before 1991, an event that now appears uncertain.

Table 9-10. *Estimated Requirements for Various Forms of Primary Energy in South Korea, 1986 and 1991*
Ten trillions of kilocalories[a]

	1986			1991		
Fuel	Electricity generation	Other	Total	Electricity generation	Other	Total
Oil	3.2	32.0	35.3	2.9	47.3	50.2
Coal	4.9	16.1	21.1	6.4	16.2	22.6
Gas	1.8	0.2	1.9	3.3	0.8	4.1
Noncommercial	0.0	1.5	1.5	0.0	0.9	0.9
Hydroelectric	0.7	0.0	0.7	0.9	0.0	0.9
Nuclear	6.6	0.0	6.6	14.6	0.0	14.6
Total	17.2	49.9	67.1	28.1	65.2	93.3

Sources: 1986 figures are from table 9-7. "Other" column for 1991 is from table 9-9. Total use of energy in generating electricity in 1991 was estimated by assuming that electricity delivered to consumers is 30 percent of energy inputs. The 1991 total for electricity generation was allocated among the various means of generation in proportion to the projected outputs by those means. See Korea Electric Power Corporation, *Electric Power in Korea, 1982* (Seoul: KEPCO, 1982), p. 24. Numbers are rounded.
a. Ten trillion kilocalories approximately equal a million t.o.e.

the increase in the share of oil in total sectoral consumption that the current five-year plan anticipates for the period 1981–86 could continue in the period 1987–91. If the projections in tables 9-5 and 9-9 are taken at face value, the share of oil in sectoral energy consumption will rise from 52 percent in 1981 to 58 percent in 1986 and to 64 percent in 1991.

As was true of the estimates for 1986 discussed earlier in this chapter, the structure of energy supply in 1991 looks quite different when account is taken of prospective changes in the ways in which electricity is generated. This is done in table 9-10. Estimates for 1986, based on the current five-year plan, are also shown for comparison. Table 9-11 converts the estimates in table 9-10 to percentages. Current plans for substituting other sources of energy for oil in generating electricity will almost counterbalance the increased importance of oil in other uses. As a consequence, the estimated share of oil in total primary energy supply is projected to rise only slightly from 1986 to 1991.

The planned increase in electric generating capacity is quite impressive. As table 9-12 shows, the Korea Electric Power Corporation intends to increase total generating capacity (excluding pumped storage) from 9,435 megawatts in 1981 to 25,404 megawatts in 1991, an expansion of almost 170 percent. Nuclear capacity will increase about 1,800 percent,

Table 9-11. *Estimated Structure of Primary Energy in South Korea, 1986 and 1991*
Percent

	1986			1991		
Fuel	Electricity generation	Other	Total	Electricity generation	Other	Total
Oil	18.7	64.3	52.6	10.3	72.5	53.8
Coal	28.8	32.4	31.5	22.8	24.8	24.2
Gas	10.2	0.3	2.8	11.7	1.2	4.4
Noncommercial	0.0	3.0	2.2	0.0	1.4	1.0
Hydroelectric	4.2	0.0	1.1	3.2	0.0	1.0
Nuclear	38.1	0.0	9.8	52.0	0.0	15.6
Total	100.0	100.0	100.0	100.0	100.0	100.0

Source: Tables 9-8 and 9-10. Numbers are rounded.

Table 9-12. *Means of Generating Electricity in South Korea, 1981 and Estimates for 1986 and 1991*

	1981		1986		1991	
Fuel	Installed capacity (megawatts)	Percent	Installed capacity (megawatts)	Percent	Installed capacity (megawatts)	Percent
Oil	7,296	77.3	4,593	27.7	3,943	15.5
Coal	750	7.9	4,030	24.3	6,030	23.7
Gas	0	0.0	1,900	11.5	2,550	10.0
Hydroelectric	802	8.5	1,282	7.7	1,665	6.6
Nuclear	587	6.2	4,766	28.8	11,216	44.2
Total	9,435	100.0	16,571	100.0	25,404	100.0

Source: Figures for installed capacity are from Korea Electric Power Corporation, *Electric Power*, p. 23. Pumped storage capacity is not included because it consumes more energy than it produces. Numbers are rounded.

coal-fired capacity about 700 percent.[27] Oil-fired capacity, however, will decrease 46 percent as several existing oil-fired plants are converted to coal or gas and no new oil-fired plants are built. There were no gas-fired plants in 1981, but plans call for 2,550 megawatts of gas-fired capacity by 1991, a figure that will constitute 10 percent of total capacity.[28]

27. The planned increase in nuclear capacity may not be achieved on schedule. Part of the program has been delayed because the demand for electricity has not increased as rapidly as was anticipated. See *Korea Herald*, September 21, 1983, p. 6.
28. As residential demand for liquefied natural gas increases during the 1990s, some of this capacity will be shifted back to oil.

Table 9-13. *South Korea's Dependence on Imported Hydrocarbons, Estimates for 1981 and Projections for 1986 and 1991*
Millions of metric tons

Fuel	1981	1986	1991
Oil	28.7	35.5	50.2
Liquefied natural gas	0.0	1.6	3.0
Coal			
Anthracite	2.8	3.4	3.2
Bituminous	7.4	14.0	16.3
Total	10.2	17.4	19.5

Sources: Oil import requirements are from tables 9-7 and 9-10. Import requirements of liquefied natural gas are from Korea Electric Power Corporation, *Electric Power*, p. 26. Import requirements for anthracite and bituminous coal in 1981 and 1986 were derived from table 1.1.2 of the appendix to the *Five-Year Energy Plan*. Import requirements for anthracite in 1991 were estimated on the conservative assumption that both total consumption and domestic production in that year would be about the same as in 1986. Import requirements for bituminous coal in 1991 are the difference between total coal requirements (see table 9-10) and total anthracite requirements in that year. Conversion factors were as follows: anthracite (imported), 0.62 t.o.e. per ton; bituminous, 0.65 t.o.e. per ton. Numbers are rounded.

Korea must import all of its oil, natural gas, and uranium and most of its coal. There is little chance that this situation will change significantly in the near future. No oil or natural gas in commercially exploitable quantities has been found. Joint Korean-Japanese exploration in disputed offshore areas between the two countries continues but has thus far been unsuccessful.[29] Korea has some uranium, but none has been produced commercially. Deposits are small, and the uranium content of the ores is low. Higher international prices of uranium would be required to justify exploitation of known deposits.[30]

Domestic coal production (all of it anthracite) has shown a tendency to level off. The five-year plan, however, calls for annual increases in coal production of 2.5 percent—to be accomplished by better financing, more realistic pricing, increased mechanization, and improved living conditions for miners and their families.[31] Whether this rate of increase in production can be sustained in the next five-year planning period is, at the least, open to question.

Estimates of Korea's import requirements for hydrocarbons can be summarized, in original units, as shown in table 9-13. Most of the

29. *Korea Newsreview*, April 11, 1981, p. 11.
30. See Argonne National Laboratory, *Republic of Korea/United States Cooperative Energy Assessment*, vol. 1: *Main Report* (Springfield, Va.: National Technical Information Service, 1981), pp. 3-43 to 3-46, for a description of uranium deposits in South Korea.
31. *Five-Year Plan*, pp. 67–74.

anthracite coal consumed in Korea, both imported and domestic, is consumed in the residential and commercial sector. A small and declining amount of anthracite is used to generate electricity. No bituminous coal was used to generate electricity in 1981, but by 1986 over 40 percent will be used for this purpose. By 1991 this percentage could be well over 50 percent.[32] Other large users of bituminous coal are the steel industry and the cement industry.

Korea plans to expand the generating capacity of its nuclear power plants from 587 megawatts (electric) [MW(e)] in 1981 to 11,217 MW(e) in 1991. All of this capacity will be in plants with light water reactors, except one 678-MW(e) plant with a heavy water reactor that was commissioned in 1983.[33] Construction of three more nuclear power plants, which would have come on line after 1991, has been deferred.[34]

On the average, every 1,000 MW(e) of light water reactor capacity uses 142 metric tons of uranium a year, and every 1,000 MW(e) of heavy water reactor capacity uses 122 metric tons of uranium a year.[35] Korea's minimum uranium requirements—all of which would have to be satisfied by imports—would therefore be roughly 663 metric tons in 1986 and 1,579 metric tons in 1991. Uranium, however, is traded in yellowcake (U_3O_8), which is measured in short tons. In short tons of U_3O_8, import requirements would be about 860 tons in 1986 and 2,050 tons in 1991. Requirements would be somewhat higher if new capacity is brought into service in those years, because the initial fuel load of a nuclear power plant is greater than average annual fuel consumption.[36] The Korea

32. It is estimated that in 1986 anthracite will supply 3.5 percent of the energy used to make electricity and that bituminous coal will provide 22.8 percent. In 1991 the share of anthracite is projected to be 1.9 percent, the share of bituminous 20.7 percent. See Korea Electric Power Corporation, *Electric Power,* p. 24. The increase from 1986 to 1991 in total energy devoted to electricity generation that is projected in table 9-10 more than makes up for the decline in the percentage share of bituminous coal, but it is not great enough to prevent a decline in the absolute amount of anthracite used in power plants.

33. Ibid., p. 23.

34. *Nuclear Engineering International,* vol. 27 (August 1982), p. 3.

35. International Nuclear Fuel Cycle Evaluation, *Fuel and Heavy Water Availability* (Vienna: International Atomic Energy Agency, 1980), p. 66. These estimates assume a plant life of 30 years and a 70 percent capacity factor. For light water reactors, which use enriched uranium, a tails assay of 0.2 percent U_{235} is also assumed.

36. In a light water reactor the initial load is about 2.5 times average annual consumption. In a heavy water reactor the initial load is 107 percent of average annual consumption. Ibid.

Electric Power Corporation estimates its requirements in 1986 at 1,086 metric tons of uranium, or 1,400 short tons of U_3O_8.[37]

In 1973 Korea obtained all of its crude oil from the Persian Gulf. In 1978 (the most recent year for which full data are available) Korea still obtained 98 percent of its oil from this same region. Over half of Korea's crude oil imports in 1978 (54.8 percent) was from Saudi Arabia. Kuwait supplied almost a third (32.5 percent). Iran accounted for 8.2 percent, and 2.5 percent came from the Saudi Arabia-Kuwait neutral zone.[38] In 1982 three-fourths of Korea's oil imports were from the Middle East.[39]

As part of its active program of resources diplomacy, the Korean government is trying to diversify the nation's sources of oil. An agreement was signed in early 1981 with the Indonesian government for joint oil exploration off the eastern coast of Java. Also in early 1981, the Nigerian government agreed in principle to sell oil to Korea, and a Korean firm concluded a two-year oil supply contract with Ecuador. Later in 1981 an agreement to import oil from Mexico was concluded in Seoul with the visiting director-general of Mexico's state-run petroleum company, Pemex.[40]

None of these arrangements appears to involve large amounts of oil. The results of the joint exploration agreement with Indonesia cannot be predicted, but it appears to be linked to a Korean effort to double the 10,000 barrels of oil supplied daily by Indonesia in 1981.[41] Korea reportedly hopes to get 30,000 barrels a day from Nigeria initially and to raise that amount to 50,000 barrels a day later. The agreements with Ecuador and Mexico were both for 20,000 barrels a day.[42] These quantities are not insignificant, but they are only a fraction of the nearly 600,000 barrels a day that Korea imported in 1981.

Comprehensive quantitative information on the sources of Korea's

37. Personal communication from Korea Electric Power Corporation, July 1982.

38. Argonne National Laboratory, *Republic of Korea/United States Cooperative Energy Assessment*, p. 3-35.

39. *Korea Herald,* May 3, 1983, p. 1.

40. These arrangements were reported in the following issues of *Korea Newsreview:* Indonesia, February 7, 1981, p. 18; Nigeria and Ecuador, February 21, 1981, pp. 8 and 16; and Mexico, August 1, 1981, p. 11.

41. In October 1982, in connection with a long-term agreement for liquefied natural gas, Indonesia agreed to raise oil exports to Korea to 15,000 barrels a day. See *Korea Herald,* October 21, 1982, p. 6.

42. Ibid.

coal imports appears not to have been published. *The Republic of Korea/ United States Cooperative Energy Assessment* states that Korea imports anthracite coal primarily from the United States, Peru, South Africa, Swaziland, and Vietnam and obtains bituminous coal mostly from the United States, Canada, Australia, Indonesia, and the Philippines.[43] The assessment gives no numbers and says nothing about Korea's substantial imports of bituminous coal from China. This coal, which is used by the Korean cement industry, is shipped directly from Chinese ports to South Korean ports under arrangements made through brokers in Hong Kong.[44] These shipments may have been suspended following the visit of Kim Il-Sun to China in September 1982.[45]

Korean resources diplomacy includes coal. Korean firms (trading companies, cement companies, the Korea Electric Power Corporation, and the Pohang Steel Company) have purchase agreements or joint ventures in several countries, including Australia, Indonesia, Canada, the Philippines, and the United States.[46] The governments of Korea and Colombia have discussed investments by Korean firms in Colombian coal mining projects.[47]

In August 1983 Korea agreed to import 2 million tons of liquefied natural gas a year from Indonesia over a period of twenty years.[48] Imports are to begin in 1986, presumably at a somewhat lower level than 2 million tons annually. By the late 1980s Korea plans to import 3 million tons a year.[49] This extra volume could come from Indonesia (as was at first anticipated by Korean planners) or from elsewhere. The possibility of obtaining liquefied natural gas from Thailand has been discussed by officials of the two governments, but no definite arrangements appear to have been made.[50]

43. Argonne National Laboratory, *Republic of Korea/United States Cooperative Energy Assessment*, p. 3-32.

44. Conversations in Seoul in November 1980 and February 1982, and the *Asian Wall Street Journal Weekly*, February 16, 1981, p. 1. These imports are seen by South Korea as both politically and economically advantageous. They have received little publicity because they are clearly objectionable to China's ally, North Korea.

45. The *Asian Wall Street Journal Weekly*, November 29, 1982, p. 9, said that the Chinese "are reported to have cut off indirect trade with South Korea under pressure from the North."

46. *Korea Newsreview*, February 21, 1981, pp. 16–18.

47. Ibid., October 10, 1981, p. 14, and *Korea Herald*, April 24, 1983, p. 6.

48. *Korea Herald*, August 16, 1983, p. 1.

49. Korea Electric Power Corporation, *Electric Power*, pp. 26–28.

50. *Korea Newsreview*, November 29, 1980, p. 17.

Korea has obtained all of its natural uranium from Canada and the United States and has relied exclusively on the U.S. Department of Energy for enrichment services.[51] This situation, however, will change in the near future. As part of the deal under which the French firm Framatome sold the Korea Electric Power Corporation two nuclear reactors, another French firm, Cogema, agreed to supply enriched uranium fuel for the reactors for a period of ten years.[52] The reactors are scheduled to begin operations in 1988 and 1989.[53] The Korea Electric Power Corporation is engaged in uranium exploration projects in Paraguay (with the Taiwan Power Company) and in Gabon.[54]

Despite the lack of full quantitative information, the preceding review of recent developments indicates that during the 1980s Korea will make considerable progress in diversifying the geographic sources of its energy imports. Most of this progress will come from the shift from oil to nuclear power and, to a lesser extent, to liquefied natural gas.[55] A series of new, relatively small arrangements with oil exporters outside the Persian Gulf will, however, reduce dependence on that volatile area to some extent.

Possible Developments after 1991

As the Korean economy matures, some slowing in the rate of economic growth is possible but by no means inevitable. Korea will still have a large agricultural sector in the 1990s, and substantial gains in productivity of land and labor will continue to be achievable by farm mechanization and the shifting of surplus farm workers to other parts of the economy. Moreover, the restructuring of manufacturing industry to emphasize more technology-intensive activities that has been begun in the current five-year plan could provide an impetus for sustained growth well beyond the end of the 1980s.

Even if a new war with North Korea and serious domestic disturbances

51. Conversations in Seoul, November 1980.
52. *Korea Newsreview*, November 15, 1980, p. 21, and *Nuclear Engineering International*, vol. 25 (December 1980), p. 3.
53. These reactors were originally to have begun operations in 1986 and 1987, but their completion has been postponed to 1988 and 1989. Personal communication from Korea Electric Power Corporation, July 1982.
54. *Korea Newsreview*, February 21, 1981, p. 17.
55. A modest increase in the use of coal is also anticipated, but its share of total primary energy supply will probably decline. See tables 9-7, 9-8, 9-10, and 9-11.

can be avoided, there will continue to be problems in South Korea that could retard economic expansion. Among such problems are conditions in Korea's export markets, the international competitiveness of Korean goods (and construction services), and the availability of both domestic and foreign capital. The balance between positive and negative factors in Korea's economic future is difficult to discern; in any case, a full analysis of the possible rate of growth of the Korean economy in the 1990s is beyond the scope of this study. For present purposes, all that need be concluded is that there are no compelling reasons for believing that a slowdown in Korea's economic growth will markedly reduce the rate of expansion in its total energy requirements.

Along with economic growth, the other determinant of total energy requirements is the overall energy intensity of the economy. Here, too, there are both favorable and unfavorable factors. Changes in the structure of Korean manufacturing industry to deemphasize energy-intensive activities, such as petrochemicals and basic materials, could work to reduce overall energy intensity. In addition, the rate of electrification of the economy could slow down somewhat, which would reduce the rate of increase in transformation losses. But improvements in thermal efficiency may be smaller than in the past as opportunities for shifting from less to more efficient fuels decline. Moreover, unless Korea can continue to borrow heavily abroad, it must achieve a favorable balance of nonfuel trade to pay for its fuel imports. When that happens, net imports of embodied energy will give way to net exports of embodied energy.

If, as appears prudent, energy requirements are assumed to continue to grow in the 1990s at a rate comparable to those projected earlier in this chapter for the 1980s, choosing the means of meeting those requirements will continue to be an important policy problem. Two interconnected questions stand out. First, how far can the electrification of the Korean economy be carried? Second, can an increase in dependence on oil be avoided?

The process of electrification may be limited by saturation of markets for electricity. The structural changes in manufacturing industry referred to above should create some new requirements for electricity, and households should use more electricity as more home appliances are acquired. But beyond these two obvious areas, opportunities for increased sales of electricity are speculative. Electrification of the railroads is one significant possibility that has already been discussed in Korean government circles. Increased penetration of the residential and commercial space-heating market by electricity is another possibility. Current thinking in Seoul, however, is that the next phase in this market will feature increased use of liquefied natural gas. Only later, when more nuclear power plants permit lower electric rates, may electricity begin to compete in the space-heating market on a large scale.

Over the long run, the market for electricity is not likely to shrink. Needs for electricity may, however, grow more slowly than in the past. As a consequence, the rate of expansion of electric generating capacity may also be reduced. The expansion of generating capacity could also be retarded by financial or staffing problems (in particular, a shortage of nuclear engineers and technicians). As of early 1982, however, officials responsible for the expansion program were confident that the needed capital could be mobilized and that skilled personnel could be trained or recruited among Koreans now employed abroad.[56]

The current five-year plan projects an impressive reduction in the share of oil in total primary energy requirements from 1981 to 1986, and the illustrative projection for 1991 shows only a small increase in the share of oil in total energy requirements. These results, however, conceal an important problem. The share of oil in activities other than electricity generation is projected to rise from 55.8 percent in 1981 to 72.5 percent in 1991 (tables 9-8 and 9-11). Over the same period the share of oil in electricity generation is projected to drop dramatically—from 77.0 percent to 10.3 percent. The possibilities of substituting other forms of energy for oil in generating electricity may therefore have been virtually exhausted by the early 1990s. Holding in check an increase in dependence on oil will then require holding down the use of oil for purposes other than generating electricity.

Prospects for success in such an effort are fair at best. Some oil might be replaced by electricity (and possibly by coal) in the industrial sector and by liquefied natural gas in the residential and commercial sector. Except for the electrification of railroads, there would appear to be little or no chance of breaking the near monopoly of oil in the transportation sector. The nonenergy sector will almost certainly continue to depend exclusively on oil.

56. Conversations in Seoul, February 1982.

In all probability Korea's dependence on oil will increase during the 1990s. Barring the discovery of oil at home, all of these rising oil requirements must be met by imports. Korean resources diplomacy may therefore become even more active as the country seeks to diversify further the geographical sources of its oil imports.

CHAPTER TEN

Major Contingencies

THE DISCUSSION thus far of Northeast Asia's future energy requirements has not considered at length contingencies that might affect the energy economies of Japan, Korea, and Taiwan. All deviations from the central case assumptions are in a sense contingencies, and the analysis has considered various alternatives to the assumed rates of change of gross domestic product (GDP) and of energy prices. The deviations that merit special attention, however, are large discontinuities that would force rapid and fundamental changes in government policies and consumer behavior.

Some conceivable technological developments would profoundly affect energy markets. In this category are the development of commercially competitive solar, fast-breeder, or fusion electric power plants and the production of lightweight, durable, and economical batteries for electric cars. But technological breakthroughs of this kind would not burst suddenly upon the world, and adjustment to them could be made gradually over a period of many years.

The contingencies that will be examined here are events that would radically alter the cost or availability of a particular form of energy over a fairly short time. The likelihood that such events will occur is not high, but neither can they be regarded as fanciful. Considering their consequences is a useful way of testing the adaptability of the Northeast Asian energy economies.

Third Oil Crisis

War or revolution could drastically reduce oil exports from the Middle East. For present purposes, the precise cause of the reduction is less important than its size and duration.[1] It will be assumed here that the reduction would be deep enough and prolonged enough to convince the countries of Northeast Asia that their dependence on Middle Eastern oil would have to be reduced as rapidly as possible. This conclusion might be reached even though the effects of the hypothetical oil crisis were less than disastrous.[2] Governments and publics alike might decide that three oil crises were too many and that past measures to reduce vulnerability to such shocks had been insufficient.

The practical question is what could be done. As shown earlier in this study, prospects for shifting to sources of oil outside the Middle East are limited. Primary reliance would have to be placed on reducing oil consumption through conservation and substituting other fuels for oil. Because its oil consumption is so much larger than that of Korea or Taiwan, attention will be focused first and principally on Japan.

The policy problem Japan would face after a third oil crisis would be the same as that after the first crisis, in 1973–74, but the circumstances would be different. Many potential savings of oil have been achieved since that crisis, and more will probably be achieved during the 1980s. The using up of opportunities for substituting other fuels for oil is especially notable.

Table 10-1 shows the sectoral changes in Japan's oil consumption from 1973 to 1979 and the further changes projected to take place by 1990 under assumptions of the central case. In 1973–79 industrial oil consumption actually fell by about 10 percent, but oil consumption in the transportation sector rose by more than a fourth. In 1979–90 industrial oil consumption is projected to increase only about a fifth as rapidly as total oil consumption, but the absolute savings are in the transformation sector, in which oil use is projected to decline approximately 20 percent. The big increases are in the transportation sector and in the residential and commercial sector, in which oil consumption is projected to increase 31 percent and 67 percent, respectively. These differences in sectoral

---

1. Developments that might cause a severe oil crisis include domestic turmoil in Saudi Arabia, war between Iran and several of the Arab countries bordering the Persian Gulf, and closing of the Strait of Hormuz at the mouth of the Gulf under any of a variety of circumstances.

2. The impact of a reduction in oil imports could be moderated for a time by drawing on emergency stocks and by reducing less essential consumption through rationing and enforced conservation. Japan (but not Taiwan and Korea) might also benefit from activation of the International Energy Agency's emergency allocation system.

Table 10-1. *Sectoral Oil Consumption in Japan in Calendar Years 1973 and 1979 and Estimated Consumption in Fiscal Year 1990*
Ten trillions of kilocalories[a]

Sector	1973	1979	1990	Rate of change, 1973–79 (percent)	Rate of change, 1979–90 (percent)
Transformation	90.6	93.2	74.7	2.9	−19.8
Industrial	94.2	85.2	87.2	−9.6	2.3
Residential and commercial	29.4	30.5	40.0	3.7	31.1
Transportation	44.4	56.5	94.2	27.3	66.7
Nonenergy uses of fuels	7.8	7.1	9.4	−9.0	32.4
Total	266.3	272.5	305.5	2.3	12.1

Sources: Figures for 1973 and 1979 are from energy balance tables pepared by the Institute of Energy Economics (Japan). Estimates for 1990 are from chapter 6 of this study. Numbers are rounded.
a. Ten trillion kilocalories approximately equal a million metric tons of oil equivalent.

Table 10-2. *Changes in Sectoral Oil Consumption in Japan in Calendar Years 1973 and 1979 and Estimated Consumption in Fiscal Year 1990*
Percent

Sector	1973	1979	1990
Transformation	34.0	34.2	24.4
Industrial	35.4	31.3	28.5
Commercial and residential	11.0	11.2	13.1
Transportation	16.7	20.7	30.8
Nonenergy uses of fuels	2.9	2.6	3.1
Total	100.0	100.0	100.0

Source: Table 10-1. Numbers are rounded.

rates of change produce the changes in the structure of oil consumption shown in table 10-2.

The figures in Table 10-2 reveal two trends that are important in the present connection. First, oil consumption is increasingly being concentrated in sectors—transportation and nonenergy uses of fuels—in which possibilities of substituting other fuels are extremely limited.[3] Second, the shares of the industrial and transformation sectors in total oil consumption have declined, or are projected to decline, markedly.

3. Nonenergy uses of fuels in Japanese statistics are entirely those of petroleum products. In 1980 oil provided 98 percent of the energy consumed in the transportation sector.

Because energy consumption in these sectors is still rising, this decline reflects the substitution of other fuels for oil. (Some substitution is also occurring in the residential and commercial sector because oil is a declining percentage of that sector's total energy consumption, even though the sector's share of total oil consumption is increasing.)

The best opportunities for accelerating the substitution process are in the transformation sector. In the central case projection of primary energy supplies in 1990 (see table 7-8), approximately 56 million tons of oil, or 18 percent of total oil supplies, were allocated to the generation of electricity. If the Japanese government were willing and able to subsidize the necessary large investments by the private utility companies, virtually all electric power plants still using oil could be converted to other fuels or could be replaced by new plants using other fuels. If such an effort were undertaken, liquefied natural gas, despite its high cost, might turn out to be the fuel of first choice. Liquefied natural gas does not cause the environmental problems that coal does, and its ability to provide peaking power efficiently gives it one advantage over nuclear energy.

In the industrial sector, the easier opportunities for substituting other fuels for oil may have been exhausted. For example, the steel and cement industries have replaced most of their oil consumption by coal. The process of substitution must depend increasingly on changes in the structure of Japanese industry. To the extent that a third oil crisis increased the cost of oil relative to other forms of energy, structural change that would reduce the use of oil would be promoted. Unfortunately the increase in oil prices would, as in past crises, probably also cause a slowdown in economic activity and retard the new investment needed to bring about structural change.

The Japanese government could most effectively achieve the goal of rapidly reducing dependence on imported oil by applying several policies simultaneously. By fully passing through to consumers the increase in oil prices caused by the crisis—and by levying special taxes on some less important uses—oil consumption could be directly discouraged (especially in the transportation sector), and switching to other fuels could be encouraged (especially in the industrial sector). The combination of an increase in oil prices and specifically targeted taxes might not be enough, however, to cause a rapid substitution of other fuels for oil. Government subsidies and preferential financing arrangements might be needed to promote structural change in industry and the virtual elimination of the use of oil in generating electricity.

The costs of such a program would be high, and the size of the reduction in oil consumption that it would achieve is uncertain. The opportunities for substituting other fuels for oil would certainly be less accessible than in the aftermath of the first oil crisis. Moreover, to the extent that a policy of reducing dependence on oil meant suppressing energy consumption rather than shifting to other fuels, the government might find that its goal of increased energy security was attainable only at the cost of a prolonged slowdown in economic activity.

Taiwan and Korea would face some of the same problems as Japan if they responded to a third oil crisis by trying to reduce rapidly their dependence on imported oil. As in Japan, the share of the transportation sector in total oil consumption is increasing in both Korea and Taiwan, and this rise has the effect of reducing opportunities to shift to other fuels. Some of the easier ways of shifting away from oil may also have been used up in both countries or will be exploited in the 1980s under central case projections.

The opportunities for an accelerated substitution of other fuels for oil may, however, be greater in the two smaller Northeast Asian economies, particularly in Korea. The oil intensity of the Japanese economy, measured by the amount of oil required to produce a unit of GDP, fell at an annual rate of 3.2 percent from 1973 to 1979 and is projected to fall at an annual rate of 3.8 percent from 1979 to 1990. In Taiwan oil intensity actually rose 3.5 percent a year from 1973 to 1979, but it is projected to decline 4.1 percent a year from 1979 to 1990. Oil intensity in Korea rose 0.7 percent annually from 1973 to 1979 and is projected to decline at an annual rate of only 0.1 percent in the period 1979–91.[4]

These figures indicate that between the first and second oil crises Japan made more progress than Taiwan or Korea in reducing the dependence of its economy on oil. They also suggest that, even without the stimulus of a third oil crisis, Japan and Taiwan may make good progress in this direction in the future. The figures raise the possibility that opportunities for substituting other fuels for oil remain relatively the best in Korea and may be second best in Taiwan.

One easily identifiable opportunity to move away from oil is in the generation of electricity. Taiwan might reduce projected oil consumption in 1990 by almost one-sixth by completely eliminating the use of oil in

4. Figures for both oil consumption and GDP (in Korea, gross national product, GNP) used in estimating changes in oil intensity were drawn from earlier chapters. The modified energy estimates announced by the Korean government in 1982 appear to call for a more rapid reduction in oil intensity.

its electric power plants. The hypothetical saving of oil in Korea (in 1991) from a similar policy would be much less—only about 6 percent of projected total oil consumption.

Nuclear Standstill

A serious accident at a nuclear power plant anywhere in the world would be regarded by many Japanese as confirmation of their deep misgivings about the safety of such installations. If the accident caused deaths or widespread radiological contamination, and if it occurred in Japan or even in nearby Korea, public reaction in Japan could bring the nation's nuclear energy program to a standstill.

Public reaction to a nuclear accident would probably be less severe in Korea or Taiwan, unless the accident occurred at a domestic power plant and was very serious. People in Korea and Taiwan are less fearful of nuclear energy than people in Japan, where attitudes are still deeply influenced by the nuclear bombing of Hiroshima and Nagasaki at the end of World War II. Moreover, the media in Japan would probably report a nuclear accident in a way that would add to public concerns, whereas the media in Korea and Taiwan would be responsive to government requests for moderation in their reporting.

In the worst case, public fears aroused by a serious nuclear accident could force the closing of existing nuclear power plants, as well as the stopping of work on new plants. Even in Japan, however, the closing of existing plants would probably be only temporary. Permanent closing would cause a prolonged electricity shortage that would disrupt the lives of millions and seriously damage the economy. A permanent shutdown would also undermine the financial stability of the electric utility companies and confront the government with the unwelcome problem of giving the utilities emergency financial assistance.

Stopping work on new plants would also be a serious blow to the utilities, but it could happen in the aftermath of a serious nuclear accident. Bringing Japan's nuclear energy program to a halt would obviously have a significant effect on the Japanese energy supply system over the long run.

Nuclear power plants in Japan generate electricity more cheaply than do conventional thermal plants. The increasing share of nuclear plants in total electricity production has tended to hold down costs and has encouraged the continued electrification of the economy. If there were to be no more additions to nuclear generating capacity, this process could be slowed down or even ended.

The most important effect of putting a cap on nuclear generating capacity in Japan, however, would be to force greater reliance on other means of making electricity. Other sources of primary electricity, hydroelectric and geothermal power, could not easily make up the shortage that would soon appear.[5] Moreover, hydropower is used in Japan to meet peak loads and could not substitute for nuclear energy as a source of base-load electricity. Most of the electricity that would have been provided in future years by additional nuclear plants would have to be generated by conventional thermal means.

For illustrative purposes, it will be assumed that, as the result of a serious accident at a nuclear power plant in 1985, Japanese nuclear generating capacity is frozen at about 23 gigawatts (capable of producing 11.5 million metric tons of oil equivalent, t.o.e.). Nuclear generating capacity had been expected to increase to 36 gigawatts in 1990, for an electricity output of 18.0 million t.o.e. About 6.5 million t.o.e. that would have been produced by nuclear power plants in 1990 would therefore have to be produced by other means.[6]

This hypothetical shortfall would amount to only 8.7 percent of estimated electricity production in Japan in 1990. If, as is likely, it were all to be made up by conventional thermal plants, the output of such plants would have to increase by about 14 percent. Much, and possibly all, of the increase could be produced by using idle capacity in existing plants. As time passed, however, the shortfall would increase, and additional conventional thermal plants would have to be built.

If a generating efficiency of 37 percent is assumed, increasing the output of thermal plants by 6.5 million t.o.e. would require an input of 17.6 million t.o.e. in hydrocarbon fuels. If this input were divided equally among oil, coal, and liquefied natural gas, imports of these fuels in 1990 would have to be increased by the following amounts: oil, 5.87 million metric tons; coal, 8.15 million metric tons; liquefied natural gas, 5.99 billion cubic meters.[7] These absolute increases in hydrocarbon imports

---

5. A study by a Japanese research organization in 1980 reportedly concluded that, if the expansion of nuclear generating capacity were suddenly to be stopped, the country would be in serious trouble within three years. (Private conversation, July 1982.)

6. See chapter 7 (table 7-3 and text) for the derivation of these estimates.

7. The following conversion ratios were used: oil, 10 trillion kilocalories per million metric tons; steam coal, 7,200 kilocalories per kilogram; liquefied natural gas, 9,800 kilocalories per cubic meter.

represent the following percentage increases: oil, 1.7; coal, 6.6; and liquefied natural gas, 13.0.[8] Such increases in imports, especially in coal and liquefied natural gas, might not be immediately attainable. There might, therefore, be a relatively brief period in which electricity production would lag behind requirements.

Over the longer run, imports of hydrocarbons would rise more rapidly in Japan than they would have if nuclear generating capacity had continued to expand. One important means of reducing dependence on Middle Eastern oil would have been lost. The resulting situation would be less than ideal, but it would nevertheless be manageable if no new oil crisis occurred.

The consequences of stopping the construction of new nuclear power plants in Korea and Taiwan—an event that is much less likely—would be similar to those in the case of Japan. The adjustments that would follow such a change in policy would be somewhat larger, however, because nuclear energy is projected to provide a larger share of total primary energy in Korea and Taiwan than in Japan.

Limits on Coal

As part of the effort to reduce dependence on imported oil, the energy policies of all three Northeast Asian countries feature the increased use of coal. The atmospheric pollution caused by burning coal, even when antipollution devices are used, is well known. This problem is one explanation for Japan's greatly expanded use of liquefied natural gas and for Korea's plans to begin importing liquefied natural gas in the mid-1980s.

The increased use of coal in Northeast Asia and elsewhere may become part of a larger problem. Some students of the world's climate believe that the burning of hydrocarbons is increasing carbon dioxide levels in the atmosphere and that, as a result, average temperatures will slowly increase. This possibility is known as the "greenhouse effect" because carbon dioxide in the atmosphere, like the glass in a greenhouse, keeps some of the energy from the sun from returning to outer space.[9]

8. Based on central case projections of hydrocarbon import requirements (table 7-16).

9. See Conservation Foundation, *State of the Environment, 1982* (Washington, D.C.: The Foundation, 1982), pp. 73–74; Sam H. Schurr and others, *Energy in America's Future: The Choices before Us* (Johns Hopkins University Press for Resources for the Future, 1979), pp. 385–88; and Electric Power Research Institute, "DOE Takes the Lead in CO_2 Research," *EPRI Journal*, vol. 7 (June 1982), pp. 34–36.

This hypothesis is not accepted by everyone.[10] If average world temperatures do increase by only a few degrees, however, major consequences, many of them extremely adverse, could follow. Melting of part of the Antarctic ice cap could raise ocean levels and flood low-lying coastal areas. Some important food-producing areas could be turned into deserts, but the agricultural productivity of other areas could be increased.

If qualified scientists conclude that the greenhouse effect is indeed taking place and that the overall consequences will be overwhelmingly adverse, the governments of major industrialized countries and advanced developing countries might join in an effort to stop or slow down the increase in the amount of carbon dioxide in the atmosphere. For present purposes, it will be assumed that these countries, including those of Northeast Asia, will agree at some future date to freeze the level of carbon dioxide emissions produced by burning hydrocarbons.

Such a freeze would not be the same as a freeze on the amount of each hydrocarbon fuel consumed. Oil produces less carbon dioxide per heat unit than does coal, and natural gas produces less than oil. Japan, Korea, and Taiwan (and other nations) would probably conform to a ceiling on carbon dioxide emissions by permitting some increase in the use of oil and gas but would balance such increases by reducing the use of coal. The hypothetical agreement to stop the increase of carbon dioxide in the atmosphere would become, in effect, an agreement to reduce the use of coal.[11]

If Japan, Korea, and Taiwan continue to follow a policy of reducing dependence on imported oil, the principal alternatives to coal would be nuclear energy and liquefied natural gas. Nuclear energy would be preferable as a means of dealing with the greenhouse effect because it emits no carbon dioxide. Under current technology, however, nuclear energy is used almost exclusively in the generation of electricity. Burning liquefied natural gas produces carbon dioxide but, as noted above,

10. The report of a panel convened by the U.S. National Research Council accepted the likelihood of a continued increase in the carbon dioxide content of the atmosphere but urged further study of this phenomenon and its consequences before deciding what actions, if any, might be required. See Carbon Dioxide Assessment Committee, *Changing Climate* (Washington, D.C.: National Academy Press, 1983).

11. Readers who are skeptical of the greenhouse hypothesis can regard the contingency raised here as simply a reversal of the policy of increasing the share of coal in primary energy supply. Such a reversal could conceivably also be brought about by an increase in the cost of using coal, including the cost of antipollution measures, relative to the cost of using other fuels.

produces much less carbon dioxide per unit of heat than burning coal.[12] Liquefied natural gas can substitute for coal in a variety of uses in addition to the generation of electricity. Neither nuclear energy nor liquefied natural gas can, of course, substitute for coking coal.

In the near term the practical problem in Northeast Asia—and worldwide—would be to reduce carbon dioxide emissions from coal enough to permit the probably unavoidable increase in emissions from oil, particularly in the transportation sector. If oil consumption levels off or declines later in the century, the problem would become one of avoiding any increase in the total use of hydrocarbon fuels. Nuclear energy would continue to provide a part of the answer, but in time its use would have to be extended from generating electricity to providing steam for heat used in industrial processes and in space heating. A steady decline in the overall energy intensity of economies would also ease the problem. Eventually newer forms of energy, such as solar and nuclear fusion, might contribute to a more lasting solution.

12. An electric power plant fueled by liquefied natural gas emits somewhat less than 60 percent as much carbon dioxide as a coal-burning plant of similar size and generating efficiency. An oil-burning plant would emit about 80 percent as much carbon dioxide as a coal-burning plant. See Schurr and others, *Energy in America's Future*, p. 386.

CHAPTER ELEVEN

Findings and Implications

THIS FINAL CHAPTER reviews the principal findings of the study and points out important implications of the possible energy developments that have been projected for 1990 and beyond.

Principal Findings

The analysis of past energy developments in Northeast Asia has produced some results that may be useful in thinking about energy problems elsewhere. It is also worthwhile to place the projections of possible future developments in regional and international contexts.

Lessons from the Past

Two lessons can be drawn from past energy developments in Northeast Asia. First, the overall energy intensity of an economy is the product of a variety of factors not all of which can be controlled by governments. Second, adjustment to large increases in energy prices—as in the oil crises of 1973–74 and 1979–80—involves trade-offs between the goals of economic growth and price stability.

ENERGY INTENSITY. The energy intensity of an economy is, of course, the average amount of energy used to produce a unit of national product. The ratio of energy use to gross domestic product, which is commonly used to measure overall energy intensity, can be thought of as the sum of sectoral ratios of energy use to GDP. As shown in the analyses of past energy developments in the Northeast Asian countries, an increase in one sectoral ratio can be wholly or partially counterbalanced by a decrease in another sectoral ratio.

Changes in sectoral ratios of energy use to GDP reflect structural changes in an economy as well as changes in the efficiency or intensity with which energy is used. Thus the energy efficiency of all modes of transportation might improve, but the ratio of energy used in transportation to GDP might still increase as the result of a shift from rail transport to less energy-efficient highway vehicles.

The country analyses in the study also show the important influence on energy intensity of several general factors: energy prices, thermal efficiency, energy lost in transforming primary energy to secondary energy, and energy embodied in international trade.

Changes in energy prices are inversely related to changes in energy intensity. As would be expected, higher energy prices have stimulated energy conservation among consumers in Northeast Asia. Even more important, higher energy prices are bringing about structural changes in the economies of the region. Less emphasis is being placed on energy-intensive activities and more on activities that are technology intensive.

Thermal efficiency, or the percentage of gross energy delivered as useful energy, is affected by changes in the relative economic importance of both fuels and consuming sectors. As the case of South Korea will illustrate, industrialization brings improvements in thermal efficiency by substituting more efficient fuels for the traditional noncommercial fuels (wood, charcoal, and crop residues) and by increasing the share of the relatively efficient industrial sector in total energy consumption. Increases in thermal efficiency to some extent counter the effect that changes in the relative importance of agriculture and industry have on energy intensity.

Transformation losses obviously contribute to energy intensity. The largest transformation losses are incurred in the generation of electricity. Industrialization usually means increased reliance on electricity, and this has been the case in Northeast Asia. In this respect, industrialization adds to energy intensity.

Governments cannot easily regulate the complex sectoral changes that determine energy intensity. Nor can they precisely control the various general factors that influence energy intensity. Nevertheless, governments do affect energy intensity—sometimes unconsciously, and sometimes by deliberately giving priority to goals other than energy conservation.

The decisions of the authorities in Taiwan and Korea in the early 1970s to expand heavy industry clearly increased energy intensity, but those decisions were justified at the time in terms of development policy. Later in the decade the emphasis on technology-intensive activities by all three Northeast Asian countries was intended to reduce energy intensity, but higher energy prices also made it good development policy. The decisions in all three countries to pass those higher prices on to energy users were also made in the interest of reduced energy intensity.

The efforts in all three countries to substitute other sources of energy for oil tended to increase energy intensity. In all sectors coal has a lower thermal efficiency than oil, and increased reliance on electricity means larger transformation losses. But in this case reduced dependence on vulnerable, imported oil was seen to be worth the cost in energy intensity.

The nonfuel portion of any country's international trade in effect embodies the energy used, directly and indirectly, in making the commodities traded. An export surplus of embodied energy pushes up energy intensity; an import surplus has the opposite effect. Japan and Taiwan illustrate the first case, Korea the second.

Japan and Taiwan have exported more embodied energy than they have imported, thereby contributing to the energy intensity of their economies. The export surpluses in these countries' nonfuel trade are consistent with government policies of using high levels of exports to sustain domestic employment and economic growth. But countries that are heavily dependent on imported energy, such as Japan and Taiwan, should be expected to have favorable balances in their nonfuel trade. Korea will someday also run a favorable balance on its nonfuel trade. It is in a phase, however, in which it is borrowing heavily to finance its development and experiences chronic unfavorable balances in both its total trade and its nonfuel trade. Korean development strategy, therefore, has the side effect of helping to hold down energy intensity.

ADJUSTMENTS TO OIL CRISES. The adjustments of the three Northeast Asian countries to the oil crises of 1973–74 and 1979–80 depended on the conditions of their economies when the crises began and on policy judgments made about the relative priorities to be given to checking inflation and maintaining economic growth.

At the time of the first oil crisis, all three countries were experiencing strong inflationary pressures. The sharp increase in oil prices therefore aggravated a problem that already existed. The authorities in Japan and Taiwan reacted by deliberately precipitating brief recessions to counter the inflationary effect of higher oil prices. The Japanese government applied restrictive monetary and fiscal policies. The government in Taipei tightened credit and reduced consumer purchasing power by increasing the prices of goods and services subject to its authority. Use of this latter technique may have been unique to Taiwan. Its success depended on persuading consumers that the increase in government-controlled prices was a one-time action and not the beginning of a new trend.

The Korean government adopted a different strategy in response to the first oil crisis. To maintain rapid economic growth, it did not try to contain the inflationary impact of the increase in oil prices. The depressing effect of higher oil prices on economic activity was reduced by passing on part of the price increase to foreign purchasers of Korean products. This was possible because of the strong competitive position of Korean products at the time.

The second oil crisis posed a somewhat different problem. The increase in oil prices was more gradual than in the first oil crisis. Governments therefore had more time to adjust, but they also were uncertain for a longer time about the magnitude of the crisis.

At the onset of the second oil crisis, prices were relatively stable in Japan and Taiwan, but inflationary pressures were relatively strong in Korea. Japan was in a particularly good position to deal with the crisis. The economy was not overheated, as it had been in 1973. Excess productive capacity existed, and labor productivity had been rising much more rapidly than labor unit costs. Structural changes in the manufacturing sector after the first oil crisis had reduced the vulnerability of the Japanese economy to further increases in energy prices.

All three governments passed the increase in oil prices on to final users more rapidly in the second oil crisis than they had done in the first crisis. The Japanese government in essence pursued a policy of reinforcing through mildly deflationary fiscal and monetary actions the economic forces that were working to moderate the impact of the increase in oil prices. No sharp choice between economic growth and price stability had to be made.

Korea and Taiwan adopted roughly similar policies in the second oil crisis. Both countries sought to reduce the inflationary effects of higher oil prices, but not to the extent that economic growth would be stopped or reversed. (Korea did experience a drop in GNP in 1980, but the economy rebounded strongly in 1981.) Both countries in effect struck a compromise between the goals of growth and stability.

In both crises all three Northeast Asian countries achieved the kind of adjustment they wanted. By different paths, they confirmed the now obvious truth: in adjusting to large increases in energy prices, stability and growth are alternatives. If more of one is desired, some of the other must be sacrificed. The experience of the Northeast Asian countries in the two oil crises also taught another lesson: the ease of adjustment to a large increase in energy prices depends on the state of an economy at the time. This lesson was particularly well illustrated by the difference in Japan's adjustments to the two crises. In 1973–74 the Japanese government had to apply the fiscal and monetary brakes hard to avoid the worsening of an already existing inflation. In 1979–80 only moderate government intervention was needed to reinforce the economy's natural adjustment processes.

Regional Energy Prospects

It is useful to review the principal findings concerning possible future energy developments in Japan, Korea, and Taiwan from both regional and comparative points of view.

PROSPECTS IN THE 1980s. Table 11-1 consolidates the central case projections of primary energy requirements of the three Northeast Asian countries in 1990. In some minor respects this consolidation is inexact. Figures for Japan are for fiscal years, and those for Taiwan and Korea are for calendar years. The small "other" category means renewable energy (largely geothermal) in Japan but noncommercial fuels in Korea. Quantitative information is not available on the use of either of these kinds of energy in Taiwan. Despite these problems, the consolidation does give a useful general picture of possible regional trends.

Total regional requirements for primary energy are projected to increase nearly 50 percent from 1980 to 1990—an average annual rate of increase of about 4 percent. The use of 1990 in this projection should not be taken too literally. Recent developments in Japan—slower economic growth and reduced energy intensity—suggest that the projected level of primary energy requirements may not be reached until sometime after 1990. This projected increase in regional energy requirements rests on central case assumptions of fairly stable energy prices and the economic

Table 11-1. *Requirements for Primary Energy in Northeast Asia,*
1980 and Estimates for 1990
Ten trillions of kilocalories[a]

	1980				1990			
Fuel	Japan	Taiwan	Korea	Total	Japan	Taiwan	Korea	Total
Oil	243.8	20.4	27.0	291.2	305.5	25.8	47.2	378.5
Coal	63.6	3.8	13.2	80.6	98.5	11.6	22.3	132.4
Gas	24.3	1.8	0.0	26.1	51.7	2.2	4.1	58.0
Nuclear	20.1	1.9	0.9	22.9	48.6	5.7	13.0	67.3
Hydroelectric	22.4	0.7	0.5	23.6	28.0	1.5	0.9	30.4
Other[b]	0.2	n.a.	2.5	2.7	1.2	n.a.	1.0	2.2
Total	374.4	28.6	44.1	447.1	533.5	46.8	88.5	668.8

Sources and notes: For Japan, table 7-8 (figures are for Japanese fiscal years). For Taiwan in 1980, Energy Committee, Ministry of Economic Affairs, *Taiwan Energy Statistics, 1980* (Taipei: The Ministry, 1981), pp. 146–47; in 1990, chapter 8 ("Means of Meeting Taiwan's Future Energy Requirements"). For Korea in 1980, Ministry of Energy and Resources, *Five-Year Energy Plan, 1982–86*, appendix, p. 3; in 1990, table 9-10 (1990 figures interpolated from 1986 and 1991 estimates). All estimates for 1990 are based on central case assumptions: fairly stable energy prices and the average annual growth rates of gross national product (GNP) or gross domestic product (GDP) specified in the text. Numbers are rounded.
    n.a. Not available.
    a. Ten trillion kilocalories approximately equal a million metric tons of oil equivalent (t.o.e.).
    b. "Other" energy is noncommercial fuels in Korea and renewable energy (largely geothermal) in Japan.

growth rates specified in earlier chapters. If, for example, energy prices increase, energy requirements will be lower than shown in table 11-1. How much lower will depend on the amount of the price increases and on the price elasticities of energy demand in each of the three Northeast Asian countries.

Energy requirements will also be different if economic growth rates vary from those assumed in the central case. Case A below assumes GDP or GNP growth rates one percentage point below those assumed in the central case. Case B assumes growth rates one percentage point above those of the central case. These alternative projections use the GDP or GNP elasticities implicit in the central case. In ten trillions of kilocalories, or approximately millions of metric tons of oil equivalent (t.o.e.), energy requirements in 1990 would be: case A, 619.7; central case, 668.8; case B, 716.0.[1]

As would be expected, Japan's energy requirements will continue to be much larger than those of Korea and Taiwan combined. The share of Japan in the regional total, however, is projected to decline (under central

1. The central case projection is from table 11-1. Cases A and B are from area projections: Japan, table 7-14; Taiwan, chapter 8 ("Energy Requirements after 1985"); and Korea, chapter 9 ("Alternatives to the Central Case"). The projections for Korea were for 1986 and 1991; the projection for 1990 was derived by interpolation.

Table 11-2. *Structure of Primary Energy Supply in Northeast Asia, 1980 and Estimates for 1990*
Percent

Fuel	1980				1990			
	Japan	Taiwan	Korea	Total	Japan	Taiwan	Korea	Total
Oil	65.1	71.3	61.2	65.1	57.3	55.1	53.3	56.6
Coal	17.0	13.3	29.9	18.0	18.5	24.8	25.2	19.8
Gas	6.5	6.3	0.0	5.8	9.7	4.7	4.6	8.7
Nuclear	5.4	6.6	2.0	5.1	9.1	12.2	14.7	10.1
Hydroelectric	6.0	2.4	1.1	5.3	5.2	3.2	1.0	4.5
Other	0.1	0.0	5.7	0.6	0.2	0.0	1.1	0.3
Total	100.0	100.0	100.0	100.0	100.0	100.0	100.0	100.0

Source: Table 11-1. Numbers are rounded.

case assumptions) from 84 percent in 1980 to 80 percent in 1990. Over the same period the share of Korea is projected ro rise from 10 percent to 13 percent. The share of Taiwan is projected to change very little, to increase from 6.4 percent to 7.0 percent.

As table 11-2 shows, substantial changes may take place during the decade in the structure of energy supply of both the region and its geographic components. The share of oil in regional energy supply is projected to decline from 65.1 percent in 1980 to 56.6 percent in 1990. Over the same period the share of nuclear energy is projected almost to double, rising from 5.1 percent to 10.1 percent of total energy supply. The shares of natural gas and coal are projected to increase by 2.9 and 1.8 percentage points, respectively.

Table 11-3 consolidates the projections of energy import requirements made in earlier chapters under assumptions of the central case. Area and regional import requirements for all fuels are estimated to be higher in 1990 than they were in 1980. Despite efforts to reduce dependence on imported oil, regional import requirements for oil rise about 30 percent. Requirements for coal rise 84 percent, and those for liquefied natural gas increase 120 percent. From a low base in 1980, import requirements for uranium are projected to increase 315 percent.

The preceding chapter considered several hypothetical contingencies that would cause the Northeast Asian countries to try to reduce or limit their dependence on particular forms of energy. Reducing dependence on oil would be more difficult than it was at the time of the first oil crisis because many opportunities for substituting other fuels for oil have already been taken. Oil consumption is increasingly being concentrated

Table 11-3. *Energy Import Requirements in Northeast Asia in 1980 and Estimates for 1990*

Fuel	1980	1990
Coal (millions of metric tons)		
Japan	73.6	123.6
Taiwan	4.6	14.6
Korea	7.2	19.1
Total	85.4	157.3
Oil (millions of metric tons)		
Japan	243.4	303.8
Taiwan	20.8	26.4
Korea	27.0	47.2
Total	291.2	377.4
Liquefied natural gas (millions of metric tons)		
Japan	22.0	45.3
Taiwan	0.0	0.0
Korea	0.0	3.0
Total	22.0	48.3
Uranium (thousands of short tons of U_3O_8)		
Japan	2.4	8.3
Taiwan	0.2	0.9
Korea	0.1	2.0
Total	2.7	11.2

Sources: Japan, tables 7-11 and 7-12 (figures are for Japanese fiscal years); Taiwan, chapter 8 ("Means of Meeting Taiwan's Future Energy Requirements"); Korea, chapter 9 ("Means of Meeting South Korea's Future Energy Requirements"); and *Five-Year Energy Plan*, appendix, p. 4. Numbers are rounded.

in sectors—transportation and nonenergy uses of fuels—in which the possibilities of substituting other fuels are quite limited.

Putting a halt to development of nuclear generating capacity would force an immediate increase in imports of hydrocarbons and the eventual construction of additional conventional thermal power plants. One important means of reducing dependence on Middle Eastern oil would have been lost, but the situation would still be manageable.

Limiting the use of coal for environmental reasons would give a new impetus to the use of nuclear energy and liquefied natural gas to generate electricity. In time the use of nuclear energy might have to be extended to provide steam for industrial process heat and for space heating.

COMPARATIVE PROSPECTS OF NORTHEAST ASIAN COUNTRIES. As table 11-1

Table 11-4. *Energy Intensity in Northeast Asia, 1980 and Estimates for 1990*

Economy	Kilocalories per constant U.S. dollar		Index (Japan = 100)	
	1980	1990	1980	1990
Japan	3,600	3,149	100	100
Taiwan	7,150	6,234	199	198
Korea	7,805	7,763	217	246

Sources: The primary energy requirements used in calculating overall energy intensities are from table 11-1. GNP figures for 1980 in national currencies at current prices are from the following sources. Japan: Organization for Economic Cooperation and Development, *OECD Economic Surveys: Japan* (Paris: OECD, July 1981), p. 70; Taiwan: Directorate-general of Budget, Accounting, and Statistics, *National Conditions* (Taipei: Statistical Bureau, Winter 1981), p. 36; Korea: Bank of Korea, *Monthly Economic Statistics*, vol. 35 (December 1981), p. 132. GNP figures in national currencies were converted to U.S. dollars by using the average exchange rates for 1980 given, for Japan and Korea, in International Monetary Fund, *International Financial Statistics, 1982* (Washington, D.C.: IMF, 1982), and, for Taiwan, in Central Bank of China, *Financial Statistics, March 1982* (Taipei: The Central Bank, 1982). GNP figures in 1990 were estimated on the basis of central case assumptions about rates of economic growth. Numbers are rounded.

indicates, the total energy requirements of the three Northeast Asian countries are projected to increase at quite different rates during the period 1980–90. Japan's requirements grow at an average annual rate of only 3.60 percent. Taiwan's requirements increase 5.05 percent annually, Korea's 7.21 percent.

These projected annual rates of increase in primary energy requirements, in conjunction with the assumed rates of economic growth, imply average income elasticities of demand for energy. For Japan this elasticity is 0.72. Taiwan's income elasticity is 0.70, and Korea's is 0.94.[2]

The projected increase in primary energy requirements and the assumed economic growth rates also imply changes in the overall energy intensities of the three Northeast Asian economies. To facilitate comparison, energy intensities of the three economies in 1980 and estimated intensities in 1990 have been expressed in kilocalories per constant U.S. dollar as shown in table 11-4. These measurements of overall energy intensity cannot be taken too literally. The 1990 intensities, of course, reflect projections of energy requirements that may prove to be off the mark. Even the relationships among 1980 intensities, which are based

2. In the central case, Japan's GDP was assumed to grow at an average annual rate of 5.0 percent from 1980 to 1990. Taiwan's GDP was assumed to increase 7.0 percent annually from 1980 to 1985 and 6.0 percent annually from 1985 to 1990. Korea's GNP increased 6.8 percent in 1981. It was assumed to grow 7.6 percent a year from 1981 to 1986 and 7.0 percent a year from 1986 to 1990.

on actual figures of energy use and GNP, may be somewhat distorted by the use of international exchange rates.[3]

Despite these problems, the rough estimates of overall energy intensity given in table 11-4 indicate the general trends implicit in the central case projections and the relative energy intensities of the three economies. If the intensity figures are taken at face value, it would appear that the overall energy intensities of the economies of Japan and Taiwan will both decline by about one-eighth from 1980 to 1990. Over the same period the energy intensity of the Korean economy will fall only about 0.5 percent.

The wide difference between the energy intensity of the Japanese economy and the energy intensities of the other two, less-developed economies is noteworthy. This difference can best be brought out by expressing the energy intensity figures as index numbers (also shown in table 11-4). The relationship between the energy intensities of Japan and Taiwan changes very little from 1980 to 1990. The gap between the energy intensity of Korea and those of the other two areas, however, widens.

The changes in the structure of energy supply projected for the three Northeast Asian countries during the 1980s display both similarities and differences. (See table 11-2.) All three countries are expected to reduce their relative dependence on oil, although their total consumption of oil is expected to increase. All three will obtain a larger proportion of their total energy supplies from nuclear power. But here the similarities end.

Taiwan is projected almost to double the share of coal in its total energy supply. The importance of coal in Japan's energy supply is expected to increase very little, and in Korea the share of coal is actually expected to decline. Natural gas is expected to become increasingly important in Japan and, in the second half of the decade, in Korea. In Taiwan, which depends on local gas and has no plans to import liquefied natural gas, the importance of gas is expected to decline.

COMPARISON OF NORTHEAST ASIAN PROSPECTS WITH PROSPECTS ELSEWHERE. The requirements of Northeast Asian countries for primary energy may grow more rapidly than those of other, comparable countries. Thus a major oil company estimated that energy demand of the high-growth developing countries (a category that includes South Korea and

3. For example, if a country's currency is overvalued in terms of the U.S. dollar, its GNP converted to dollars will be too high, and its overall energy intensity will be too low.

Taiwan) will increase at an average annual rate of 4.6 percent during the period 1980–2000.[4] This rate compares with the central case growth rate in the present study of 7.2 percent a year for Korea and 5.0 percent a year for Taiwan for the period 1980–90.

The same source estimated that energy demand in the major industrialized countries would grow at the following annual rates from 1980 to 2000: Western Europe, 1.5 percent; United States, 1.0 percent; and Japan, 2.0 percent.[5] The central case of the present study projects a 3.6 percent annual increase in Japan's primary energy requirements over the decade 1980–90. The difference between the two estimates for Japan can in large part be explained by differing assumptions about economic growth rates and changes in energy prices. Even with these differences, however, the oil company projected a more rapid growth of energy consumption in Japan than in Western Europe or the United States.

Prospects after 1990

Projecting the total primary energy requirements of the three Northeast Asian countries in the 1990s would be a highly speculative exercise. It may be useful, however, to examine in a nonquantitative manner the principal variables that may determine those requirements.

Whether the three economies will grow as rapidly in the 1990s as was assumed in the central case for the 1980s is uncertain. The shift to more technology-intensive activities that, in varying degree, is already under way could continue and could help to sustain economic growth. In Korea—but not in Japan, and only to a lesser extent in Taiwan—a substantial impetus to growth could also still be provided by transferring labor from agriculture to other sectors of the economy. Positive domestic influences in all three economies could be overcome by adverse developments in their export markets. The availability of foreign capital will

4. Exxon Corporation, *World Energy Outlook* (New York: Exxon, 1981), p. 38. This estimate assumes that the real price of Middle Eastern crude oil will increase 50 percent from 1980 to 2000 (about 2.05 percent annually). Economic growth assumptions are not stated in this source.

5. Ibid., pp. 30–35. The assumption about the real price of Middle Eastern oil was given in footnote 4. GNP was assumed to increase at the following annual rates: Western Europe and the United States, 2.5 percent; Japan, 4.0 percent. Again, the central case of the present study assumes fairly steady real energy prices and an annual growth rate of 5.0 percent in Japan's GDP.

continue to be important to sustained economic growth in Korea and Taiwan.

Changes in overall energy intensity could, of course, either reinforce or counteract the effect of changes in economic growth rates on primary energy requirements. Continued rapid economic growth would facilitate the purchase of more energy-efficient equipment and would cause energy intensity to decrease. The increased emphasis on technology-intensive activities should also favor a decrease in energy intensity. A slowdown in the rate of electrification would work in the same direction by reducing the rate of increase in transformation losses. Such a slowdown could be caused by the saturation of markets for electricity.

Other possible developments could cause energy intensity to decline more slowly or even to rise. These include greater use of air conditioning (in Taiwan and Japan), the spread of private car ownership (in Taiwan and Korea), slower gains or losses in thermal efficiency by increasing use of coal (possibly in all three), and shifting from net imports to net exports of embodied energy (in Korea).

The combined effect of changes in economic growth and energy intensity on primary energy requirements in the 1990s cannot be foreseen. A continued increase in those requirements, however, appears probable. If energy intensity declines—an event that is by no means assured—the rate of decline is unlikely to be so great as to cancel the positive effect of economic growth on energy requirements.

The move away from heavy reliance on oil that was stimulated by the 1973–74 oil crisis and reinforced by the second oil crisis in 1979–80 may slow down in the 1990s. Opportunities to substitute other fuels for oil—especially in the generation of electricity—will be gradually exhausted. Commercial incentives to use other fuels will weaken if, as appears possible, the price difference between oil and such fuels narrows. Moreover, as dependence on Middle Eastern oil is reduced, policy arguments to reduce the use of oil could lose some of their force.

Implications of Future Prospects

The findings of this study concerning possible future energy developments in Northeast Asia have implications both for international trade and for international relations in general.

Table 11-5. *Share of Northeast Asia in Global Primary Energy Consumption, 1980 and Estimates for 1990*
Ten trillions of kilocalories[a]

Fuel	1980			1990		
	World	North-east Asia	North-east Asian share (percent)	World	North-east Asia	North-east Asian share (percent)
Oil	2,542	291	11.4	2,618	379	14.5
Coal	944	81	8.6	1,368	132	9.6
Gas	889	26	2.9	1,012	58	5.7
Other	531	50	9.4	952	99	10.4
Total	4,908	447	9.1	5,950	669	11.2

Sources: World figures exclude Communist areas. World figures for 1980 are from U.S. Department of Energy, Energy Information Administration, *1980 International Energy Annual* (Washington, D.C.: DOE, September 1981, pp. 6–12; those for 1990 are from DOE, Energy Information Administration, *1981 Annual Report to Congress, Synopsis,* p. 43. Figures for Northeast Asia are from table 11-1. World figures for 1980 are preliminary estimates of energy production, which equal energy consumption if stocks do not change. World figures for 1990 are the midprice scenario of the source cited. Department of Energy figures in quadrillion British thermal units (Btu) were multiplied by 25 to convert them to millions of t.o.e. Numbers are rounded.
a. Ten trillion kilocalories approximately equal a million t.o.e.

International Trade

The share of Northeast Asia in total global energy consumption (excluding Communist areas) may increase significantly in the 1980s. Table 11-5 gives actual energy consumption in 1980 and estimated energy consumption in 1990 for both Northeast Asia and the non-Communist world. The figures for Northeast Asia are from table 11-1; the world figures are from publications of the U.S. Department of Energy (the world estimates for 1990 are the midprice projections).

In 1980 Northeast Asia accounted for over 9 percent of primary energy consumption outside Communist areas. If the projections in table 11-5 are taken at face value, in 1990 Northeast Asia's share of world energy consumption will rise to over 11 percent. The region's share of coal consumption will increase from 8.6 percent to 9.6 percent, and its share of oil consumption will go up from 11.4 percent to 14.5 percent. The largest change is in Northeast Asia's share of gas consumption, which is projected to rise from 2.9 percent to 5.7 percent. Because Northeast Asia must import most of its oil, coal, and gas,[6] it would play an even

6. Coal reserves are limited, and oil and gas reserves (in Japan and Taiwan) are extremely small. Efforts to find more substantial deposits of oil and gas have thus far been unsuccessful.

greater role in energy import markets in 1990 than is suggested by these shares of total consumption.

Northeast Asia's share of the world uranium market will probably increase even more than its shares of the markets for hydrocarbon fuels. A study by the Organization for Economic Cooperation and Development (OECD) estimated annual uranium requirements outside Communist areas at 30,000 metric tons in 1980 and 53,000 to 65,000 metric tons in 1990. If these figures are converted to short tons of yellowcake (in which uranium is traded internationally) and if the midpoint of the 1990 estimate is used, requirements would be 39,000 short tons of U_3O_8 in 1980 and 76,700 short tons in 1990.[7] On the basis of the estimates in table 11-3, the share of Northeast Asia in world uranium requirements would more than double during the 1980s, increasing from 6.9 percent in 1980 to 14.6 percent in 1990.

All three of the Northeast Asian countries are projected to increase their shares in world consumption of both primary energy and uranium from 1980 to 1990. Japan will continue to be much more important in international energy markets than Korea and Taiwan combined. The two smaller economies are, however, projected to record relatively greater increases than Japan in their shares of both energy consumption and uranium requirements.

The growing importance of Northeast Asia in the world energy economy means that developments there have increasing weight in influencing trends in energy markets. But beyond this obvious fact, the relative dynamism of the Northeast Asian part of international markets for energy materials manifests itself in many ways. Northeast Asian firms and government agencies will become even more involved in energy exploration and development projects in other parts of the world than they are now. Similarly, producers of energy-related materials and services in many countries will seek markets in Northeast Asia.

The United States should have increased opportunities to sell coal to Northeast Asia. If the prohibition in U.S. law is removed, Alaskan oil could displace some Middle Eastern oil in Northeast Asian markets. Over the longer run, exports of liquefied natural gas from Alaska to Japan may increase, and new markets for Alaskan liquefied natural gas may be found in Korea and Taiwan. The United States may also be able

7. Organization for Economic Cooperation and Development, Nuclear Energy Agency, *Nuclear Energy and Its Fuel Cycle: Prospects to 2025* (Paris: OECD, 1982), p. 19. One metric ton of uranium equals 1.3 short tons of U_3O_8.

to take advantage of growing Northeast Asian markets for equipment, technology, and services used in oil exploration and development and in nuclear energy. In development of nuclear power, however, European and Japanese competition will be strong.

In recent years the United States has been able to increase its exports of basic petrochemicals and aluminum ingots to Northeast Asia, because U.S. producers have had access to cheaper feedstocks (petrochemicals) or electricity (aluminum). At least in basic petrochemicals, this U.S. trade advantage may prove to be transitory. By 1990 the United States may be a large net importer of ethylene and ammonia and its derivatives. The major exporters of ethylene in 1990 may be Canada, the Middle East, Western Europe, and the USSR. Principal exporters of ammonia in 1990 may be the USSR, Mexico, Trinidad, and the Middle East.[8] Changes in patterns of trade in these and other energy-intensive commodities have, of course, been caused by the large increase in energy prices during the 1970s. At first, higher energy prices affected the competitive positions of existing producers. Later, higher energy prices caused shifts in the location of facilities producing energy-intensive items. Such facilities tend increasingly to be located close to cheap sources of energy.

The likelihood that the Northeast Asian countries will import increasing amounts of energy in future years has wider implications for international trade. To pay for rising energy imports, these countries must either borrow abroad or generate increasing export surpluses in their nonfuel trade. In the past, Korea has taken the first course, and Japan and Taiwan have taken the second. In the future, all three countries will probably try to pay for their energy imports with their nonfuel exports. In running export surpluses in their nonfuel trade, the Northeast Asian countries inevitably compete with industrialized countries, including the United States, that export similar commodities. This will continue to be the case, even if Northeast Asia's total trade—and Japan's trade in particular—is in balance.

The nuclear energy programs of the Northeast Asian countries also pose commercial problems for the United States and other industrialized countries. Japanese competition in the sale of nuclear equipment and services has been mentioned. Increased use of electricity generated by nuclear power plants may also give the Northeast Asian countries a

8. *Oil and Gas Journal*, vol. 81 (May 9, 1983), p. 84.

modest competitive advantage. In Northeast Asia—contrary to recent U.S. experience—nuclear electricity is cheap electricity. As more nuclear electricity becomes available, the costs of some of Northeast Asia's exports will be reduced.

International Relations

The efforts of the Northeast Asian countries to increase the reliability of their supplies of imported energy, described earlier in this study, can be expected to continue. By entering into long-term energy supply arrangements and by investing in energy and other projects, the ties between the Northeast Asian countries and nations exporting various energy materials will in general be strengthened. South Korea and Taiwan will reap some benefits useful in their continuing political struggle against their Communist adversaries. Japan will in some cases be able to reduce troublesome, favorable balances in bilateral trade.

Increased activity by Northeast Asian countries in energy-exporting countries will also create some problems. Disputes over prices are inevitable, and local resentment over the appropriation of energy resources for foreigners may be encountered. The net effect of the resource diplomacy of the Northeast Asian countries is, however, likely to be strongly to their advantage.

Some tension between the energy goals of the Northeast Asian countries and other foreign policy goals will undoubtedly persist. Japan's requirements for imported oil will increase, and prospects for obtaining these requirements outside the Middle East are limited. Japan will therefore experience chronic difficulties in balancing its need for uninterrupted supplies of Middle Eastern oil against its strong interest in good relations with the United States. Each new Arab-Israeli crisis will find Japan trying to separate itself from the pro-Israeli aspects of U.S. policy but being restrained by a desire to avoid unduly offending its powerful ally.

Similar problems may occasionally arise in Japan's relations with the Soviet Union, whenever a Japanese interest in developing Siberian energy resources runs afoul of U.S. economic sanctions against the USSR. Even the continuing disagreement between the United States and Japan over the adequacy of Japan's defense effort has an energy dimension. A part of the U.S. case for stronger Japanese defense forces is the argument that Japan should share in the task of deterring military threats to Japan's oil supply lines from the Middle East and Southeast Asia.

For Taiwan and Korea, opportunities to diversify sources of energy are limited to some extent by political considerations. Peking and Moscow might be willing to export energy materials to Taiwan, and might even see a political advantage in doing so. Taipei will, however, probably continue to regard such trade as contrary to its anti-Communist principles.

Perhaps the most significant energy development in Northeast Asia is the continuing shift from primary dependence on Middle Eastern oil to dependence on a variety of energy sources from Pacific countries: coal from Australia, Canada, and the United States; uranium from Canada and Australia; enrichment services from the United States; liquefied natural gas from Southeast Asia and Alaska; and oil from Southeast Asia. Energy supplies from other Pacific countries will probably also increase—for example, coal from China, Siberia, and Colombia; liquefied natural gas from Siberia and Australia; and oil from Mexico.

A pattern of energy trade is emerging that adds substance to the concept of a Pacific Basin community. Whether this concept can and should be given concrete institutional form is a subject extending far beyond the scope of the present study. In energy, however, the time for the exchange of views among Pacific countries on common problems has clearly arrived. Actual cooperation on specific problems may lie some distance ahead.

Cooperation among energy-importing Pacific countries in an emergency is conceivable. One can imagine, for example, a Northeast Asian oil stockpile to supplement national supplies. The need for regional cooperation in an energy emergency has not, however, been widely recognized. In addition, practical problems, including the relationship between a regional scheme and the emergency plan of the OECD's International Energy Agency, remain to be addressed.

Although serious obstacles have to be overcome, the most likely area for regional cooperation is in nuclear energy. Japan and the United States are discussing the possibility of further cooperation in the development of fast breeder reactors. Japan and the United States have also discussed the possibility of establishing a storage facility for spent nuclear fuel on a Pacific island, but agreement actually to undertake this project is at best uncertain.

The expansion of civil nuclear energy facilities in Northeast Asia could create opportunities for regional cooperation in that part of the Pacific Basin. It also has implications for U.S. nuclear nonproliferation policy.

Cooperation in nuclear energy among the Northeast Asian countries, and possibly among other countries of the Western Pacific, could take a variety of forms, including the exchange of operating information, mutual assistance in emergencies, joint investments in fuel supply projects, and collaboration in research. The political impetus for cooperative arrangements, however, is not yet strong. More ambitious endeavors, such as a regional reprocessing plant, are probably only ideas for the more distant future.

U.S. nonproliferation policy should try to anticipate the problems that will arise as the expanding Northeast Asian nuclear programs enter new phases. Problems of nuclear waste management will become more acute. The United States is more or less reconciled to the reprocessing of spent fuel of U.S. origin from Japanese nuclear power plants, either in Japan or in Western Europe. Pressure from Korea and Taiwan for similar treatment is predictable. U.S. policy must also prepare for the likely accumulation of stocks of spent nuclear fuel in Northeast Asia. As more spent fuel is reprocessed, adequate controls over separated plutonium must be devised.

Note on Technical Terms and Methods of Analysis

ENERGY economics uses technical terms and analytical methods that may not be familiar to nonspecialists. The following explanations of these terms and methods will help readers to follow the analyses of past and possible future energy developments in Northeast Asia that are presented in this study.

Energy Balances

Energy balances are used to present quantitative summaries of the supply and disposition of energy in a given country for a given time, usually a calendar or fiscal year. To the extent possible, the energy balances in this book follow the format developed by the Organization for Economic Cooperation and Development. Energy balances for Japan conform closely to the OECD model. It proved impossible, however, to construct energy balances for South Korea and Taiwan that conform to the OECD model in every respect. Energy balances present three kinds of information: primary energy supply, transformation losses, and final energy consumption.

Primary energy supply equals domestic production of various fuels, plus net imports (or minus net exports) of those fuels, plus withdrawals from fuel stocks (or minus additions to stocks). Primary energy supply, in this context, includes imported secondary fuels, such as coke and refined petroleum products. Primary energy supply also includes hypothetical inputs associated with the generation of primary electricity; that is, electricity generated by hydro-, nuclear, or geothermal power. These hypothetical inputs equal the caloric value of the fuel that would be needed to generate the same amount of electricity by conventional thermal means.

Transformation losses (sometimes also treated as an energy-consuming sector of the economy, the transformation sector) are the net inputs of energy used in generating electricity, refining oil, making coke and briquets from coal, and making town gas from oil and coal derivatives. Transformation losses include both losses inherent in transformation processes and the energy used by the facilities (such as power plants and refineries) engaged in these processes. The hypothetical inputs in making primary electricity are regarded as transformation losses. These inputs thus appear in the energy balance twice: as positive components of primary energy supply and as negative components of transformation losses.

Final energy consumption equals primary energy supply minus transformation losses. Final energy consumption includes both the untransformed part of primary energy supply (for example, coal consumed as coal) and the forms of energy produced by various transformation processes (for example, electricity). Final energy consumption is broken down by energy-consuming sectors, which are not the same as those used to calculate sectoral contributions to gross domestic product (GDP); most commonly these energy-use sectors are industrial, residential and commercial, transportation, and nonenergy uses of fuels (usually oil).

The industrial sector usually includes energy consumed by agriculture as well as by manufacturing. The residential and commercial sector includes miscellaneous energy consumption and is sometimes (as in OECD usage) referred to simply as the "other" sector. The nonenergy-uses sector should in principle include fuels used as chemical feedstocks, but in practice those fuels are usually recorded in the industrial sector. The nonenergy-uses sector is then limited to a few relatively small uses of oil, such as in asphalt and lubricants.

Energy balances could be expressed in any of a variety of units. In this book, they are expressed in kilocalories, which are easily converted to metric tons of oil equivalent (t.o.e.). Ten billion kilocalories approximately equal a thousand t.o.e.; ten trillion kilocalories approximately equal a million t.o.e. Kilocalories are used in most of the tables, and t.o.e. are in general used in the text.

Gross and Useful Energy

Energy data in heat units, such as kilocalories, give gross energy content; that is, the amount of energy that theoretically could be obtained if a given form of energy were used with 100 percent efficiency. Except in nonenergy uses, however, complete efficiency is not attainable. Thermal efficiency, or the percentage of gross energy delivered to final consumers as useful energy, varies with both the kind of energy and the way in which the energy is used. In general, electricity (at the point of delivery) has the highest thermal efficiency, followed by gas, liquid, and solid fuels in that order. Among principal energy-consuming sectors, the industrial sector tends to be the most efficient; the residential and commercial sector follows in efficiency, and the transportation sector is last.

Changes in the structure of energy supply or in the structure of energy use affect the overall thermal efficiency of an economy. Structural changes in energy supply or use can therefore increase or decrease the consumption of useful energy, even though the consumption of gross energy remains the same. Identifying shifts in the consumption of useful energy is important to a full understanding of how an economy uses its energy supplies.

Obtaining accurate coefficients of thermal efficiency for each fuel in each kind of use is very difficult. The principal coefficients used in this study were derived by William D. Nordhaus on the basis of experience in the United States and Western Europe.[1] Coefficients for wood and charcoal, used only in calculations for the residential and commercial sector, are estimates by Resources for the Future.[2] In contrast with the coefficients developed by Nordhaus, those for wood and charcoal reflect the experience of developing countries. The coefficients of thermal efficiency used in this study, by fuel for three principal energy-consuming sectors (with "n.a." meaning "not available"), are:

1. William D. Nordhaus, ed., *Proceedings of the Workshop on Energy Demand* (Laxenburg, Austria: International Institute for Applied Systems Analysis, 1975), p. 527.
2. Personal communication from Joy Dunkerley (Resources for the Future, Washington, D.C.).

	Residential-commercial	*Transportation*	*Industry (except energy)*
Solid	0.20	0.044	0.70
Liquid	0.60	0.220	0.80
Gas	0.70	0.220	0.85
Electricity	0.95	0.400	0.99
Wood	0.10	n.a.	n.a.
Charcoal	0.30	n.a.	n.a.

No such coefficients appear to have been developed on the basis of the experience of any of the three Northeast Asian countries examined in this study. Moreover, any single set of coefficients cannot identify the effect of technological changes on thermal efficiency. The estimates of consumption of useful energy and of overall thermal efficiency presented are therefore at best rough approximations. They do, however, throw light on one important aspect of the energy developments in the region.

Energy Embodied in Trade

Countries import and export energy in two forms: as fuels, such as oil and coal, and as energy embodied in other commodities. Embodied energy is the energy used, directly and indirectly, in making a given commodity. Imports of embodied energy reduce fuel requirements; exports of embodied energy increase fuel requirements.

The magnitude of net imports or exports of embodied energy depends on the balance of a country's nonfuel international trade and the energy intensities of the commodities entering into that trade. The Energy Research Group at the University of Illinois at Urbana-Champaign has calculated energy intensity coefficients for 355 sectors in the input-output model of the U.S. economy constructed by the Bureau of Economic Analysis (BEA) of the U.S. Department of Commerce.[3] The sectors used by the BEA, however, do not match in detail the standard international trade classification, or SITC, used in international trade statistics. It was therefore necessary in this study to use six broad categories in applying energy intensity coefficients to trade data: food,

3. Bruce Hannon, Robert A. Herendeen, and Thomas Blazeck, "Energy and Labor Intensities for 1972," ERG doc. 307 (University of Illinois at Urbana–Champaign, Energy Research Group, April 1981).

beverage, and tobacco; crude materials; chemicals; manufactured goods; machinery; and miscellaneous.

Since the energy coefficients are based on 1972 U.S. data, some distortions could result if—as has been done in this study—the coefficients of energy intensity are applied to trade data for other countries in other years. They do, however, make it possible to estimate the rough magnitude of energy embodied in trade.

Structure-Intensity Analysis

The overall energy intensity of an economy is measured by the ratio of total energy use to GDP or to gross national product (GNP). This ratio is the sum of sectoral ratios of energy use to GDP; thus changes in overall intensity can be viewed as the net result of changes in the sectoral ratios. Whether changes in the sectoral ratios are the result of structural changes in the economy, changes in the sectoral intensity of energy use, or both is of some interest. The structure-intensity analysis used in this study is an attempt to separate structural and intensity effects on sectoral ratios of energy use to GDP. The procedure followed is similar to that described by Joel Darmstadter, Joy Dunkerley, and Jack Alterman.[4]

In structure-intensity analysis, structural and intensity factors are changed separately to determine their effect on the sectoral ratios of energy use to GDP. The sum of structural and intensity effects will, however, be less than the change in the sectoral ratio being analyzed. The difference, known as the "interaction effect," arises because the relation between the structural effect and the intensity effect is multiplicative rather than additive.

Income and Price Elasticities of Demand

Two other economic measurements, as applied to energy consumption, merit definition. The income elasticity of energy demand is the percentage change in energy use associated with a given percentage change in GDP. The price elasticity of energy demand is the percentage change in energy use associated with a given percentage change in energy prices.

4. Joel Darmstadter, Joy Dunkerley, and Jack Alterman, *How Industrial Societies Use Energy: A Comparative Analysis* (Johns Hopkins University Press for Resources for the Future, 1977), pp. 252–58.

Index